RADIO CONTROL AIRPLANE
Workshop Secrets

Contents

6 Introduction

CHAPTER 1
Building Basics

8 Choosing wood for better modeling
by George Wilson Jr.

12 Building strong, lightweight structures
by George Wilson Jr.

15 Aligning and balancing your model
by George Wilson Jr.

19 Building strong, light wings
by George Wilson Jr.

24 How to strip-plank a fuselage
by John Tanzer

27 How to inlay balsa sheeting
by Randy Randolph

28 Make built-up truss ribs
by Graeme Mears

32 Sanding basics
by Mariano Alfafara

36 Sheet foam wings with plain brown paper
by Bertil Klintbom

38 Make laminated paper parts
by Harry B. Cordes

41 Mount wings using differential screws
by Al Ehrenfels

42 Work with EPP foam
by David M. Sanders

47 Working with metal
by George Wilson Jr.

CHAPTER 2
Landing-Gear Essentials

52 Landing-gear ABCs
by George Wilson Jr.

56 Ski and float landing gear
by George Wilson Jr.

60 Install landing gear in a foam wing
by Les Morrow

63 Install shock-absorbing Oleo landing gear
by Gerry Yarrish

65 Make custom landing-gear clips
by Randy Randolph

66 Make concealed axles
by C. H. Bennett

68 Make a steerable tailwheel
by Ron Bozzonetti

69 Another approach to building a steerable tailwheel
by William R. Nielsen Jr.

72 Make giant-scale skis
by Roy Vaillancourt

75 Make soda bottle snow skis
by Elson Shields

78 Make carbon fiber landing gear
by Thayer Syme

CHAPTER 3
Control-Linkage Setups

82 Easy Z-bends
by Randy Randolph

83 Make sewn hinges
by Bob Underwood

85 Install direct-control aileron servos
by Gerry Yarrish

87 Install Robart hinges in ARC models
by Gerry Yarrish

89 Control-linkage setup
by Gerry Yarrish

92 Control linkages
by Mike McConville

94 Control linkages for giant-scale models
by Mike McConville

97 Stop control-surface flutter
by Mike McConville

99 Commonsense control linkages
by Greg Hahn

102 Servo-operated Fowler flaps
by Bob Almes

107 Assemble a servo power connector
by Faye Stilley

110 Control-surface loads
by Mike Leasure

112 Add-on canards
by Roy L. Clough Jr.

CHAPTER 4
All About Engines

114 Build a small-engine test stand
by Randy Randolph

115 Build a ½A starter
by Roy L. Clough Jr.

116 How to make a 12- or 24-volt starter box
by Faye Stilley

118 Drill and tap engine mounts
by Gerry Yarrish

121 Make a recessed engine firewall
by Gerry Yarrish

124 Make a concealed muffler
by Tony Newsom

126 Part 1: Engine noise—problems and solutions
by Tore Paulsen

134 Part 2: More engine-quieting solutions
by Tore Paulsen

CHAPTER 5
Using Tools and Jigs

142 Organize your workshop
by Jim Sandquist

145 Drill bit tips
by Gerry Yarrish

148 How to use a micrometer and calipers
by Jim Sandquist

149 The perfect balance
by George Wilson Jr.

151 Get the CG right
by Roy Day

155 Build a jig for Great Planes' Slot Machine
by John Tanzer

156 Make an inexpensive wire soldering jig
by Joe Beshar

158 Enlarge 3 views
by Randy Randolph

159 Enlarge rib sections
by Randy Randolph

160 Easy vacuum-forming
by Syd Kelland

163 Build an inexpensive field stand
by John Gorham

CHAPTER 6
Building Projects

166 Build and install scale cockpits
by Gerry Yarrish

170 Make scale cockpit details
by Gerry Yarrish

173 Add power and RC to Robart's foamie F-16
by Nick Ziroli Sr.

176 Convert a rubber-powered model to a ½A glow
by Randy Randolph

178 Convert a rubber-powered free-flight to electric RC
by Tom Hunt

183 Convert a small glow plane to electric
by Joe Beshar

185 Renovate a retired flyer
by Henry Haffke

189 Add plastic interplane struts
by Roy L. Clough Jr.

Introduction

Become a better modeler with "Workshop Secrets," one of the "Master Modeler" series of books. This volume is a collection of some of the very best building how-to's that have been published in *Model Airplane News* magazine. Written by the pro's, these articles cover several aspects of airplane construction, from selecting the correct wood for your project to basic metalworking to advanced engine installation and setup. You'll find that these photo-illustrated tips and techniques are easy to understand and apply in your own projects.

Whether you're building a trainer kit or a scratch-built design, this 194-page book has the step-by-step solutions to your modeling questions. If you're looking for a new challenge, check out the chapter on building projects, including how to make your own vacuum-former and how to convert lightweight free-flight models to electric RC.

From basic construction techniques to advanced ways to quiet your engine, "Workshop Secrets" is a valuable resource that you—and your models—shouldn't be without.

Group Editor-in-Chief Tom Atwood
Editors Gerry Yarrish, Debra Sharp
Assistant Editor Bob Hastings
Copy Director Lynne Sewell
Senior Copyeditor Molly O'Byrne
Copyeditor Corey Weber
Corporate Art Director Betty Nero
Assistant Art Director Victoria Howell
Staff Photographer Walter Sidas
Director of Operations David Bowers
Production Associate Tom Hurley
Director of Circulation Ned Bixler
Circulation Assistant P.J. Uva

Chairman Aldo DeFrancesco
President and CEO L.V. DeFrancesco
Senior Vice President Yvonne M. DeFrancesco
Vice President G.E. DeFrancesco

PRINTED IN THE USA

AirAGE
www.airage.com

Copyright 2000© by Air Age Inc.   ISBN: 0-911295-52-6

All rights reserved, including the right of reproduction in whole or part or any form. This book, or parts thereof, may not be reproduced or transmitted in any form by any means without the consent of the publisher.

Published by Air Age Inc., 100 East Ridge, Ridgefield, CT 06877-4606 USA; (203) 431-9000; fax (203) 431-3000; www.airage.com

Building basics

CHAPTER 1

Choosing wood for better modeling
by George Wilson Jr.

As scratch-builders, one of our primary concerns is which wood we should choose for particular parts. Woods vary with respect to strength, weight and how easy they are to work with and bend. Heavier woods are stronger, but they weigh more, so we have to decide which characteristic is more important for a particular part and choose the wood accordingly.

This article applies to RC models of medium size, but most of the information given can be applied to models of all sizes.

Balsa, spruce, birch (dowels), lite-ply and aircraft-grade (hardwood) plywood are all readily available from your hobby shop and mail-order suppliers, and they'll satisfy all your model-building needs. In other countries, obechi and basswood are popular, and you'll see them used in foreign kits. Pine (a soft wood) can also be useful: I occasionally use it for internal parts that require strength and little finishing, because it's relatively strong, fairly light and inexpensive. (You'll need a table saw and a fine blade to cut it to size.)

BALSA
Because this is the traditional model-building wood, most of this section discusses its features and uses, and I compare other woods with it. It comes in different weights and grains, both of which are important. The skill lies in selecting the correct type for a particular application.

It comes in strips, sheets and blocks, and scratch-builders frequently cut strips off sheets with one of the many available stripping tools.

Sheet balsa should be selected according to its intended use. It bends most easily with the grain. To bend it around compound curves, cut it into thin strips and apply it as planking. To make it conform to complex or very pronounced curves, e.g., when making laminated wingtips, or even the curvature in the top forward wing sheeting, first soak it in water or, even better, a 50:50 solution of water and ammonia. Then glue the wet balsa with a water-soluble glue (typi-cally, an aliphatic-resin type), and allow it to dry thoroughly.

Features
- Light but relatively strong.
- Easily cut and shaped with simple bench tools.
- Readily available in a wide range of sizes and weights.
- Easy to bend—an important considera-

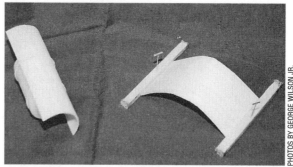

Balsa bends most easily along its grain. After being soaked in a 50:50 solution of ammonia and water, it bends fairly easily. (Use a water-soluble glue on damp wood.) The photo shows bends going with the grain and across it. Note the cracks in the cross-grain bend (the wood samples were cut from the same sheet).

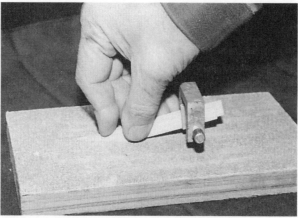

A typical collection of sheet and stick woods bought for scratch-building. Balsa, spruce, hardwood plywood and lite-ply are included; they'll become part of an OV-10 Bronco built from *Model Airplane News* plans.

Here balsa is being tapered for a scarf joint. The two ends are clamped together to ensure an equal angle on each. A belt or disk sander can also be used for this operation.

tion in model building.
- Easy to dent and crush—a unique strength feature—so it will absorb energy in a crash, and this will limit damage incurred by expensive parts such as the radio and engine.
- Easy to repair. You can remove dents from unfinished balsa by wetting the dented area. Repair damaged areas by cutting them out and gluing new balsa into the voids. At the field, you can often pull the broken balsa back into shape and glue it with epoxy or CA.

Weight
Comes in weights of from 4 to 16 pounds per cubic foot.
- Light. Use the lightest for indoor models; I have successfully used it for wing and tail sheeting on light RC models.
- Medium. Ten- to 12-pound balsa is the most readily available and is most useful in models of med-ium size.
- Heavy. The heaviest balsa is hard to find in model shops.

Grain
See Figure 1. Choose balsa with long, straight grain

Examples of A-grain (straight, long) and B-grain (straight, but shorter) balsa used for flat and curved wing sheeting.

lines; it's least likely to split when bent, and it bends easily along its grain. In its catalog, Sig Mfg. shows pictures of the various grains and says what each is best used for. When selecting balsa, this catalog makes a good starting point.
- A-grain—light to medium; used for curved surfaces, such as wing and tail sheeting and capstrips.
- B-grain—medium to very hard; widely used for ribs, bulkheads, tail structures and wingtips.
- C-grain—very hard; used for flat surfaces such as wing leading edges and fuselage longerons.

Glues and finishes
When it's coated with dope, balsa becomes much harder but with minimal weight increase. If you plan to finish it with an iron-on covering, there are coatings available to ensure good adhesion. After finish-sanding, I apply a couple of coats of nitrate dope (butyrate is acceptable) thinned 50 percent and then the prep coat—typically, Balsarite or Sig Stix-It.

Balsa basics
- Like most woods, when it's wet, balsa tends to warp, and it also does so when it has been cut into strips. This is caused by its internal stresses being relieved. To counteract this, wet your balsa and hold it bent in the opposite direction until it has dried. Usually, an equal and opposite bend does the trick.
- The edges of sheets that must be butted against each other (as when joining sheets for wing sheeting) should first be cut straight with a straightedge of the appropriate length—a real straightedge (most

The ABCs of balsa

A-grain sheet balsa has long fibers that show up as long grain lines. It's very flexible across the sheet and bends around curves easily; it also warps easily.

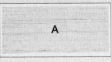

B-grain sheet balsa has some of the qualities of both A- and C-grain balsa. Its grain lines are shorter than those on A-grain, and it feels stiffer across the sheet. This is a general-purpose sheet balsa.

C-grain sheet balsa has a beautiful mottled appearance, is very stiff across the width of the sheet and splits easily. Used properly, it helps to build the strongest, lightest models. It is most resistant to warping.

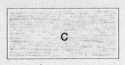

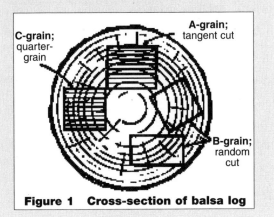

Figure 1 Cross-section of balsa log

CHAPTER 1

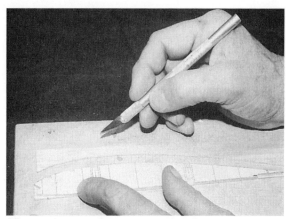

Ribs can be cut using a pattern, as shown in this photo. Although I've tried other ways of making ribs, I prefer this old-fashioned approach. Rib blanks can be stacked and then cut on a band saw or jigsaw. (Make sure the blade is at right angles to the table.) A rib-shaped block may be cut on a saw, and then the ribs may be sliced off with a saw. A friend of mine uses a router to produce smooth, accurate ribs from a pattern.

area is several times the width of the stick.

Butt joints are unreliable and should be avoided. If you must use one, reinforce it with a strong scab. To repair a break, make a good butt joint at the broken edges by wetting them thoroughly with glue; then push them together and hold them until the glue has dried. For scabs, spruce—the next wood I'll talk about—is more suitable.

SPRUCE

Spruce sticks are readily available in sizes from $1/16$ to $3/8$ inch thick. Spruce is most often cut with a razor saw, and it should be spliced in the same way as balsa. Aliphatic-resin glue, which dries slowly, is recommended because of the time required for this dense wood to absorb the glue. Sheet spruce is not normally available.

yardsticks are not truly straight). Bear in mind that if you wet the sheet after you've straightened its edge, it may curve again.

• When sheeting a wing, avoid making butt joints against wing leading edges. The structure should be designed to allow the sheeting to overlap the edge; trim it after the glue has dried. The

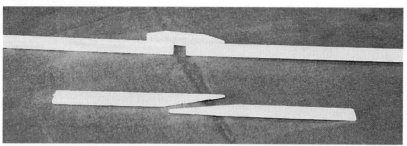

Here, balsa sticks are being prepared for joining. The top stick will have a scab to reinforce the butt joint. The lower joint is a scarf or tapered joint; this is the preferred method of joining both stick and sheet wood. Good joints in wood are at least as strong as a continuous piece.

rear edge of a wing's forward sheeting does not have to be straight if it butts up against the rear capstrips.

• If you have a sheet in which there's a long, grainwise split, you can repair it with aliphatic resin or CA. Glue the split balsa over wax paper, and pin it carefully to eliminate the gap. If it is glued to the structure (ribs or formers), the split won't be detectable, and strength will not be compromised.

• Most scratch-builders are frugal (read: cheap?), so they splice scrap balsa pieces together to make pieces in the lengths and sizes they want. After they've been sanded and finished, well-made joints are invisible, and they are as strong as the unspliced material. Splices work best when they are placed over part of the structure, e.g., a wing rib or a fuselage upright. The diagonal scarf joint is recommended for both sticks and sheets.

When you work with sheets, use a scarf angle of about 45 degrees. Scabs are permitted (a secondary piece glued across the joint to strengthen it). In a stick, the angled scarf cut should be long enough to ensure that the length of the overlapping joined

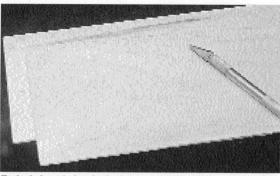

Typical sheet balsa; both pieces appear to be B-grain (random cut). This type of wood is widely used in model building. The stain on the bottom piece is no threat to its usefulness. A-grain was cut with the tree rings and has long, straight grain lines. C-grain has a mottled grain and is stiff across the sheet. It splits easily because it was cut across the tree rings.

FEATURES

• It's denser, so it's about twice as heavy as medium-weight balsa, but it's much harder and, therefore, stronger. The Sig catalog claims spruce is 10 times stronger than balsa, but it doesn't define which type of strength is being assessed.

• Not easy to bend. My simple test comparison of

samples of ¼x⅛-inch balsa and spruce indicated that it took five times as much pressure to bend spruce as it did balsa.
• Resists being crushed.

Uses
• Most useful for wing spars and fuselage longerons. These can be made with thinner, lighter spruce sticks than if you were using balsa. A ¹⁄₁₆x¼-inch spruce spar may be used in place of a ¼-inch square, medium-weight balsa spar, and it will weigh only half as much.
• Great for scabs.

BIRCH
The dowels we use are most often made of birch, which is strong and relatively heavy. Birch dowels are used for wing tie-downs and to position removable parts of models.

Like spruce, birch should be glued with a slow-drying glue that will be absorbed by the wood; poorly glued, badly fitted dowels come loose—as we all know from experience! To ensure that the rubber bands do not slip off the wing-tie-down dowels when the wood is coated with fuel residue, the dowels should project an inch or more beyond the fuselage.

POPLAR
Lite-ply is most often ⅛ inch thick and made of three plies of poplar. Lite-ply sheet is stronger than balsa, and it weighs about twice as much as medium balsa. It can be cut fairly easily with a hobby knife, but a motor-driven jigsaw is the tool of choice.

When used for fuselage sides, lite-ply is much less easily dented than balsa. That it's more difficult to die-cut is evident in the crunched edges that show up on the lite-ply parts in some kits. Lite-ply is recommended for fuselage doublers (inside) where it adds strength without a serious weight penalty.

HARDWOOD PLYWOOD
This has great strength, but it's heavy, so it's most useful where great strength and abrasion resistance are needed. It comes in three, four and five plies. One sample of 3-ply, ¹⁄₃₂-inch-thick plywood weighed 53 pounds per cubic foot—about five times the weight of medium balsa. Aircraft-grade plywood is readily available in thicknesses of from ¹⁄₆₄ to ¼ inch; the ¹⁄₆₄-inch wing-skin material has three plies.

Typically, plywood is used for the firewall that supports a model's engine and also to support landing gear. It may be drilled accurately to take close-fitting machine screws and rubber engine-vibration isolators.

Like spruce and birch, hardwood plywood is best attached with a slow-drying glue such as aliphatic-resin or thinned epoxy, both of which are fuelproof and will soak into the plywood layers.

How often have you heard: "What size should my wing spars and longerons be?" and "What's the best thickness for the wing or fuselage sheeting?" Most often, answers to these questions will be based on experience gained from building previous designs. Most of us don't have the talent, time, or resources to pick wood sizes scientifically. As a result, most models are over-designed from the strength standpoint (assuming that they have not been designed specifically for maximum crash-resistance).

The most common in-air failure is the collapse of the dihedral wing joint because of a violent maneuver—intentional or not. I have designed and flown trainer wings with structures so light that I felt they would fail, but they didn't. To select wood for your model, my advice is to try the wood dimensions that are used in similar models that fly well.

In the following section, I discuss structural designs that maximize strength and minimize weight. Until then, keep building.

CHAPTER 1

Building strong lightweight structures

By George Wilson Jr.

The author's Seasquare, built in the early 1960s, was very strong. Of the three built initially, one is still being flown by the author; the others may also still be in use. The wing uses LE and TE D-tube construction, and the tail surfaces are diagonally braced. This type of braced construction uses less wood than flat-sheet construction, but it's stronger because of the varying grain directions. The hull is balsa box construction. This model has survived many crashes.

A nice feature of scratch-building is that you can use your favorite construction methods. There is no perfect way. Perhaps you planned a lightweight model, or maybe an aerobatic type that needs strong wings. In any case, well-built structures and appropriate outer surfaces will provide the necessary strength without an undue increase in weight. It's inconceivable, however, to think that a structure can be built within allowable weight limits that will withstand the "ultimate" crash. Further, bear in mind that heavier models fly faster and crash harder.

FACTORS OF STRENGTH

Factors that help determine a model's strength are the design of the structural members (spars and frameworks) and the use of outer surfaces ("stressed skin") to distribute and absorb stresses created in flight and on the ground, even in moderate crashes. Glues and glue joints also influence strength.

DESIGN TIPS

• Strong wings. There are many good techniques for building wings, and a detailed discussion of wing construction will have to wait for another column. I usually prefer to use plywood dihedral braces, but many designers like the "stressed-skin" approach and use a fiberglass covering over the wing's center. In any case, in most RC wings, the main spar should be a variation of an "I-beam" (see Figure 1). Of the variations shown, my favorite is "D," the full-depth spar type where the ribs are in two pieces. The parts of these ribs fit efficiently on the sheet wood so you'll have more ribs per sheet. The leading edge (LE) sheeting ("D-tube") section forms the upper and lower parts of the I-beam. Note that rib notches aren't required, and that simplifies building. This successful type of construction provides lengthwise strength with minimum weight. The I-beam spar and stressed skin with the D-tube LE are strong enough to survive when twisted. Incidentally, D-tube construction at the trailing edge (TE) as well as the LE provides additional bending and twisting strength. A

Figure 1. Cross-sections of spars that work well are shown. "A" is the classic I-beam with its web centered between the top and bottom members. This type of beam is used in many full-size structures, such as bridges. "B" is common variation of the I-beam using an off-center web. Vertical-grain webbing is most often used, but this writer has used horizontal-grain webbing to increase the wing's lengthwise bending strength. "C" is a "box beam"—an I-beam variation with two webs. "D" is a variation that uses D-tube construction as part of the I-beam.

Typical "double D-tube" wing construction is shown in this photo. Both the LE and TE have D-tube construction, and the spar and LE D-tube form an I-beam. Note that the spar is full depth; the ribs are in two pieces. The dihedral brace is a substantial piece of plywood. The center top sheeting was added later, and the center was covered with fiberglass and resin for added stressed-skin strength.

BUILDING BASICS

wing with LE and TE D-tube is very rigid and relatively light. It will be even stronger if a strong covering is used over both the open areas and over the sheeted areas (more on this later).

• Strong fuselage. A framework like the time-honored stick-fuselage construction can be figuratively developed: start with a solid block of wood and remove material (and weight) until it is like a section of diagonally braced fuselage.

Figure 1 shows the development of a stick fuselage from a block. If you are curious, build an unbraced section ("C" in the figure) and push on its corners. It will probably break easily. Brace it diagonally (F) and push the corners again; the difference is great because the brace distributes the reacting forces, as shown in "F."

GLUE CHOICES

Part "H" of the figure shows a neat corner joint that should be well glued to ensure that it is strong and able to transfer forces. I prefer aliphatic-resin glue (Titebond, Sigbond, etc.) and, anyway, I'm strongly allergic to cyanoacrylate. For strong joints, use aliphatic-resin glue and add small glue fillets to the corners of the joints.

CAs are rigid adhesives that can be brittle if they have been cured too quickly—typically, when an accelerant is used—or improperly formulated. Brittle joints are generally a result of improper glue use. Thin CA will readily be wicked into the wood's cracks and pores, especially balsa's. To guarantee that there is enough glue in the joint, you'll have to apply the glue more than once. If you've been modeling a while, I'm sure you'll remember the need to double-glue joints when you used Ambroid cellulose glue, or when you used an older-style wood glue on end grain—different techniques for different glues. CA simply requires different techniques to be effective.

STRONG JOINTS

To ensure strength, many builders overuse glue, but neatly fitting joints are far more important. Sloppy joints bridged with glue do not provide strength, and they may increase weight. On the other hand, after a joint has been made, the use of small, neat fillets of glue around butt joints will add strength but minimal weight. The fillets act as "mini-gussets" and are similar to the full-scale gussets used in steel and wood construction.

Hobby outlets offer a choice of miter boxes that will help you make good, repeatable joints. All these devices allow you to cut many sticks at exactly the same angle; some boxes feature adjustable

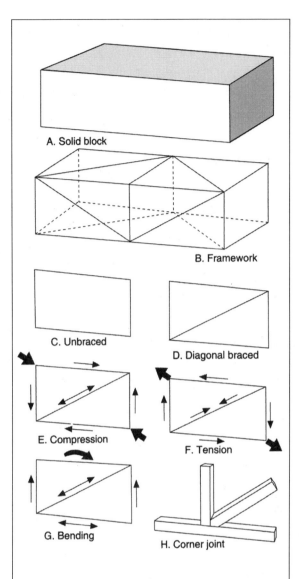

Figure 2. The figurative development of the diagonally braced framework (typically, a fuselage) is shown. The block is cut away until all that is left is the stick structure. The open framework (B) is surprisingly strong when compared with the block. The bracing accounts for much of the strength. An unbraced section (C) is very weak. The addition of a diagonal brace (D) makes the section quite strong. In parts "E" through "G" of this figure, the heavy arrows represent external forces, and the light arrows represent the reaction forces generated in the framework to counteract the external forces. In compression (E), the brace is in tension. In tension (F), the brace is in compression. In both cases, the brace strengthens the frame. In bending (G), the framework transfers the reaction forces to the sides and the diagonal. To ensure the best transfer of the forces within a frame, the corner joints must fit well (H) and be well glued.

angles. The ultimate tool for angling the ends of sticks is a sandpaper block. Cut the stick slightly longer than you need, then sand it to fit.

CHAPTER 1

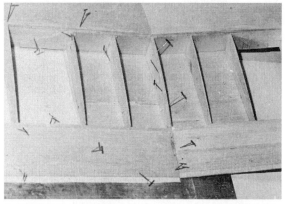

A typical balsa box fuselage before the top and bottom sheeting were added. Note that it is being built over a centerline with the bulkhead locations shown on it. To distribute the stresses along the fuselage sides, gussets are used at stress points. Lite-ply has become popular for building box fuselages. The author prefers balsa; it's lighter and easier to work with.

COVERINGS

The subject of covering materials that contribute structural strength is another long one . (For an in-depth review of the subject, see "Stressed-Skin Design" by Andy Lennon. It is complicated by arguments about appearance, ease of use, sagging over time, brittleness and built-in adhesives. "Stressed skin" may be of covering materials such as plastic (Micafilm, MonoKote, etc.) or fabric (silk, Koverall, etc.), or it may be wood, such as balsa or plywood. For a convincing demonstration, build a wing framework but don't cover it; then flex it. Next add D-tube sheeting to the top and bottom LE, apply the overall covering and flex the wing again. The increase in stiffness will be dramatic.

My preference is for the less shiny types of covering because I find them more realistic. On the other hand, most aircraft seen at our flying fields (except scale models) are not truly realistic in design or appearance.

Carl Goldberg was fond of quoting from the 1939 book by French author Antoine de Saint-Exupéry, "Wind, Sand and Stars." Saint-Exupéry was a pilot and a philosopher who flew the early African routes. Carl's favorite part of Saint-Exupéry's book was his statement that "perfection is finally attained ... when there is nothing left to be taken away ...." Carl would add, "So it

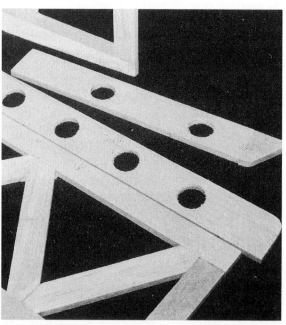

Typical diagonally braced tail surfaces. This type of construction is lighter than sheet surfaces and is stronger because of the varying grain directions. It is also more warp resistant. Note the neat fits at the corners: these ensure strength.

is with model airplanes." Incidentally, "Wind, Sand and Stars" is still available in my local library; it is well worth reading.

Frameworks should be neatly fitted. Sloppy corners, as shown in this photo, may lead to failures and crashes. Aliphatic-resin glue (see text) is preferred. Apply a little around the joints to form gussets.

BUILDING BASICS

Aligning and balancing your model
by George Wilson Jr.

Whether you're a kit builder or a scratch-builder, your model has to be balanced and aligned as the plan calls for if the model is to fly well. At a "Discover Flying" RC club presentation, Mario Borgatti discussed the importance of aligning and balancing models. With Mario's permission, I've borrowed some of the aligning techniques he discussed to share with you.

NECESSARY TOOLS

Alignment tools can be inexpensive, but they produce accurate results. Alignment involves angles—mostly right angles that can be measured using drafting or carpenter's triangles; for best accuracy, use the largest one available. Other angles are measured using protractors or incidence meters. Incidence meters are used to set wing and stabilizer (stab) incidences, which are the angles of the wing and stab with respect to the model's horizontal reference line. Randy Randolph discusses how to build an incidence meter in his book, "RC Airplane Building Techniques" (published by Air Age Inc.).

Levels are used in many phases of aligning and balancing models. Line levels and surface levels cost around $2. A plastic 360-degree dial level costs less than $10, can be compared with a good carpenter's level and can be sanded to improve its accuracy, if necessary.

A level stand is a piece of ¼x½-inch medium or hard strip balsa that's mounted vertically on about a 3-inch-square base. It's used to elevate the aft end of the fuselage while you set the incidence angles.

You can easily make handy triangle tools for setting the fin's angle with respect to the stab (see Figure 1). These tools can also be used to square bulkheads, fuselage sides, ribs, etc., and you can mount fins using the triangle tools that don't have right-angle corners.

ALIGNING THE WING

First, transfer the fuselage horizontal reference line from the plan onto both

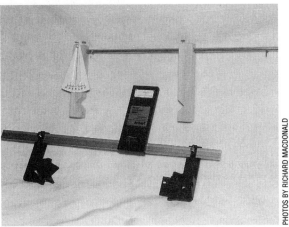

Here are scratch-built and commercial (Robart) incidence meters. This device allows precise setting of the wing and stab incidences.

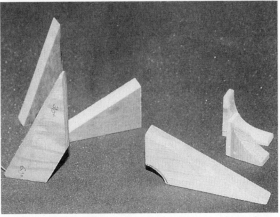

Several scratch-made alignment triangles are shown here. Figure 1 explains how to make them. Commercial versions of this sort of triangle are also available. The missing corner is very useful when you're setting a fin at a right angle to the stab.

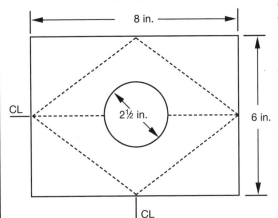

Figure 1. Eight handy triangles can be cut out of a 6x8-inch (or larger) piece of ¾-inch-thick wood. First, mark the centerlines, then cut a round hole in the center (the exact size isn't important). The missing corners in the four center squares fit around the fuselage when the square is next to the fin. Next, cut along the centerlines, and cut along the diagonals. Check the triangles out with a reliable square, sand them, and coat them with shellac, dope, or a similar sealer.

CHAPTER 1

sides of the fuselage. To avoid denting the wood, use a very soft pencil, or mark the line with masking tape. Next, mark the fuselage centerline (CL) on the top and bottom. The line should go between the centers of the firewall and the tail post. Use a good straightedge, and ignore the bulkhead center marks; they should be very close to the CLs just marked.

The wing(s) must be at right angles (perpendicular) to the fuselage CL and at the correct incidence angle. Although rubber bands work well and are time-honored means of attaching wings on trainer and sport models, I recommend that you slightly modify your model to use bolts instead.

The bolt-on system uses one or two dowels in the LE that mate with holes in a piece of plywood in the fuselage. You may not be able to mount the front dowel(s) in the wing's LE of some models, such as those with high wings. Instead, a dowel mount can be built under the wing as was done on the Live Wire Champion shown in the photo. I've also modified a Lazy Bee this way; both models fly well.

At the TE, the wing is secured with two nylon bolts (one on very small models) that should shear off in a crash. My .40-size and smaller models use 10-32 or smaller bolts in contrast to the ¼-20 used most often. Even my 10-32 bolts have survived crashes that would have been much less destructive had the bolts broken. The alignment instructions that follow assume that you use bolts—not rubber bands—to attach your model's wings.

To set the wing incidence, first find which incidence is called for on the plan. The angular setting should be shown; if it isn't, draw a line on the plan that connects the center of the TE and the most forward point of the LE. This is the usual reference for setting wing incidence. An incidence meter will

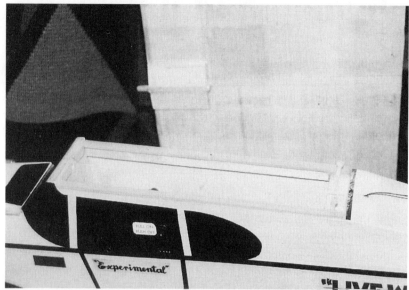

The bolt-on wing modification on a recently built Live Wire Champion. Note the dowel mount installed under the LE of the wing. The hold-down bolts are 10-32 nylon. They're strong but should shear off in a crash to reduce overall damage.

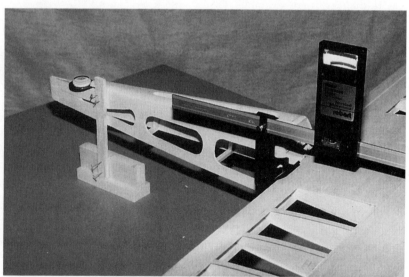

An incidence meter is used to set the wing incidence. The angle is taken from the plan. The incidence meter automatically sets itself to the wing's reference line (the mean chord) from the LE to TE.

automatically select this line as the wing's reference line. Using a protractor, measure the (incidence) angle with respect to the fuselage reference line. Put the LE dowel(s) into its socket, and pin the wing in place. Be sure the wing is at a right angle to the fuselage CL (an eyeball check is sufficient). Set the airplane on a level work surface, and raise the aft end using a level stand. When the reference line on the fuselage is level, pin the tail of the plane into place on the level stand. Install the incidence meter on the center of the wing, and check the wing's incidence.

If the angle isn't correct, adjust the wing saddle as necessary, and make sure that the fuselage is upright

BUILDING BASICS

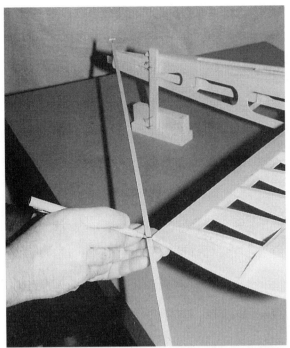

The wing is aligned to be at a right angle to the fuselage CL. Using long pieces of stripwood and measuring to the wingtip ensures accuracy.

Setting the stab incidence and aligning it with the fuselage. Usually, the stab is on the horizontal reference line or parallel to it; the incidence meter is not needed in this case.

equal, then pin the wing firmly in place.

The next step is to make the wing hold-down bolt holes. I'll assume that you've already firmly mounted a 3/16- or 1/4-inch piece of aircraft plywood in the fuselage to receive the bolts. Mark the hole positions on the wing, and drill through the wing perpendicular to the wing's surface and on through into the plywood in the fuselage. Repeat on the other side. The drill should be the bolt's tap drill size: no. 29 (9/64 inch) for 8-32, no. 21 (5/32 inch) for 10-32 and no. 7 (13/64 inch) for 1/4-20. Saturate the holes with thin CA. Tap the holes in the plywood, and enlarge the holes in the wing to accept the bolts using no. 18 (11/64 inch) for 8-32, no. 9 (3/16 inch) for 10-32 and 1/4 inch for 1/4-20. Resaturate the holes with CA, and redrill and retap to remove any excess CA. Install the bolts with washers under their heads. The wing now has the correct incidence angle and is at a right angle to the fuselage CL.

ALIGNING THE STAB AND FIN

Aligning the stab is essentially the same procedure as used for the wing. First, ensure that the incidence angle is correct and, using a bubble level, that the stabilizer saddle is horizontal. Most often, the stab is on or parallel to the horizontal reference line. Shim or trim as necessary. Use an incidence meter if the stab is not supposed to be parallel to the reference line.

Pin the stab in place. Set it at right angles to the fuselage CL as you did the wing using a point on the fuselage CL as far forward as practical as the reference and points on the LE tips as the measuring points.

The stab can now be covered and glued into place. To ensure a strong glue joint, remove the covering where the stab and fuselage will be joined. Here's a tip: before you cover the model, outline the areas you don't want to be covered with masking tape. Then cover the top and bottom—including the masked areas. Slit the covering at the edge of the masking tape and peel off the tape—neat and easy! Using this method, you can also allow for the triangular fillets frequently used between the stab and fin and the fuselage.

The fin must be on the fuselage CL and be at a

and doesn't lean sideways. Use a bubble level across the fuselage to check that the wing is horizontal.

Now make sure that the wing is perpendicular to the fuselage CL. Using a straight piece of 1/4-inch-square balsa pinned to the center of the tail post, mark the distance to a point on one TE near the wingtip. Swing the strip to the same point on the other TE, and determine whether the distances are equal. Move the wing until both distances are

WORKSHOP SECRETS 17

CHAPTER 1

The fin is easily centered using two long, straight balsa strips that tie its alignment to a point as far forward on the fuselage CL as possible.

The fin is set at a right angle to the stab using a triangle that's missing one corner. This allows the triage to fit around the corner of the fuselage and/or the fillet that may be used between the fin and the stab or fuselage.

right angle to the stab. First, trial-fit the fin, and sand as necessary to make it fit well. Pin it in place on the CL at the tail post. Pin the ends of two long, straight pieces of ¼-inch-square balsa on each side of the fin. Bring the other ends together on the fuselage CL as far forward as possible. Now pin the front of the fin in place, and make a line on the center of the fin's LE and the fuselage. The mark on the fuselage should be on the CL.

ENGINE THRUST

Many plans call for the engine thrust line (the direction of the propeller shaft) to be offset. Most often, the propeller (shaft) will point down (to offset increased lift at higher speeds) and to the right (to offset the propeller's "torque"). (The term "torque" is not strictly correct in this case, but suffice to say, most models tend to turn left without the addition of right thrust.)

If you use an engine mount that's attached to the firewall, the thrust line adjustment is best made with a hardwood shim or washers between the mount and the firewall. If the engine is mounted on bearers or on a shear plate on top of the bearers, the downthrust adjustment can be added using washers under the rear engine-mounting tab holes. Right thrust can be achieved by making the mounting-bolt holes a bit oversize and twisting the motor sideways. You can build side thrust into the shear plate when you cut the opening in it for the engine. Be sure to use locknuts on the engine-mounting bolts.

Engine thrust adjustments aren't critical in trainer/sport models; an eyeball check of them using a protractor will usually be sufficient.

FINAL NOTES

The landing gear and wheels should be at right angles to the fuselage. Use the same method as you used to align the wing and stab. To ensure good tracking while taxiing, the wheels should be set at a slight toe-in angle (pointing slightly inward).

The final test of balance and alignment comes during the initial flights. Minor warps and uneven drag (typically because of a muffler) may require you to trim the model's control surfaces. Ideally, for minimum drag, the fin and stab mountings should be adjusted to allow the fin and stab to be in line with the rudder and elevator. This is probably overkill for trainer and sport models, but it's essential for aerobatic models.

CONCLUSION

Don't overdo the accuracy of the incidence and alignment settings; it's OK to need a bit of control trim on trainer and sport models. On aerobatic, pattern and other contest models, however, the total removal of warps and accurate alignment are essential to ensure that the models will fly well at all speeds.

And, as Mario said during his presentation, "If you build it like the plan, it will fly as the original did."

BUILDING BASICS

Building strong, light wings
by George Wilson Jr.

The wings on my sorta-scale Consolidated PT-1 have solid tips, D-tube LEs and solid TEs. Without interplane struts, the wings performed very well for the draggy PT-1. Its engine is exposed for maintenance (just as in the full-scale aircraft).

Most scratch-builders have favorite types of wing construction, and most are like me: we never build a wing that's exactly the same as any previous one. This section addresses the construction of sport and trainer wings for .10- to .60-size models. Wings are possibly the most important part of a model airplane from a flight-performance standpoint. Therefore, give them a lot of thought. The principles covered apply to larger and smaller models.

SOME POINTS TO CONSIDER
- Select wood sizes to match a model's size.
- Use lighter stock. Most wings are stronger than they need to be.
- Obtain strength by proper design, not by using oversize lumber.

There are two principal strength concerns in wing design: first, bending spanwise, and second, twisting chordwise. Neither of these is forbidden, as we see gliders being towed up with wings curved because of the pressure on their undersides. Similarly, free-flight-endurance model wings twist as the model spirals upward and fights to gain alti-

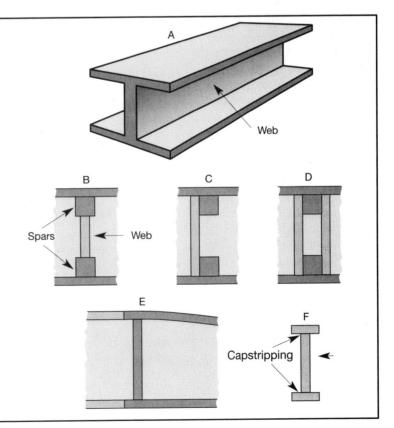

FIGURE 1. I-BEAM SPARS

A shows a classical metal I-beam used to construct buildings, bridges, cranes and airplanes. B through D are cross-sections of typical built-up spars with webbing added between the ribs. B is similar to the classical I-beam. C and D are easier to construct using webs attached to the sides of the top and bottom spars. With webs on both sides, the spars becomes a "box beam" as shown in D. The top and bottom spars are often made of spruce when extra strength is needed. E uses a full-depth spar. The ribs are made in two pieces, and the wing sheeting acts as the top and bottom members of the I-beam. This design is very strong because the beam has a continuous web and the top and bottom are far apart. F shows a cross-section of a capstripped rib. It, too, has an I-beam form.

GETTING STARTED IN RADIO CONTROL AIRPLANES 19

CHAPTER 1

tude. Well-designed wings accept this type of distortion without fracturing. Most sport/trainer models are designed with relatively rigid wings. This simplifies their design and construction.

WING CONSIDERATIONS

• **Strength.** This is the most important construction feature. It can be achieved without undue weight by using I-beam spars and D-tube (stressed-skin) construction. (It is called "D-tube" because the wing's front cross-section is shaped like the letter D.) These construction techniques take many forms. In fact, even the wing covering contributes to the I-beam nature of many spars.

• **Stress.** In a wing spar, stress is maximum at its center and tapers to zero at the wingtip. Also, if the wing's planform or thickness varies, stress is not uniform along the spar's length. Ideally, a wing spar may be tapered toward the wingtip to save weight, but construction becomes difficult. Bear in mind that most in-flight wing failures are ruptures at or near the dihedral joint (or joints, in polyhedral wings).

• **Weight.** This is always a prime consideration: light models always fly best. Good design can achieve strength with lightness. The fact that sheet wood can be bent easily with its grain and with difficulty across its grain is the key to strong wing design. Typically, thin plywood braces add great strength at the dihedral joints. Ribs are strong chordwise but weak spanwise. Luckily, they need little strength spanwise.

• **Building ease.** This is another consideration. As model designs have progressed over the years, they have become simpler and easier to build; so it is with wings. The most frequently used technique is D-tube, which is easy and produces strong, relatively light wings (see Figures 2 and 3).

A large, double-D-tube wing under construction. Note that the forward webbing is pieced from scrap balsa. After the top TE sheeting had been added, installation of the capstrips completed the wing.

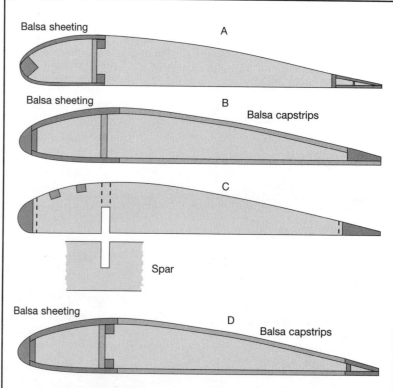

FIGURE 2. RIB LAYOUTS

This shows four of the many possible rib layouts. A is similar to those found in many kits in which the ribs are die-cut or laser-cut. It uses typical D-tube construction at both the LE and TE and, as a result, is very strong. B shows a type of construction that is strong and easily scratch-built. The ribs are in two pieces and the spar is full-depth. The combination of the full-depth spar and the LE sheeting forms an I-beam with a continuous web. C is a lightweight simple design that was once very popular—so-called "egg-crate" construction. The spar and ribs are notched to fit into one another. Notched LEs and TEs are recommended. Use of turbulator spars will make this type of wing more resistant to twisting forces. D shows a strong D-tube design that uses an upper and lower spar that can be of spruce if extra strength is necessary.

BUILDING BASICS

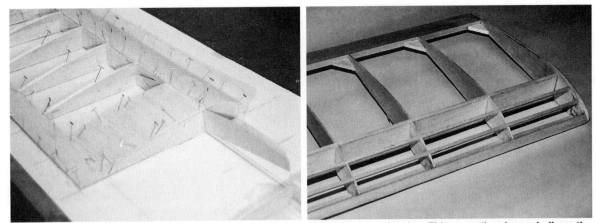

Left: this wing panel is being built over a set of bare-minimum lines, not over the plan. This saves the plan and allows the scratch-builder to make changes. The dihedral brace is 1/16-inch plywood that is very strong. The wing is double D-tube (both LE and TE). If you look carefully, you will see that the ribs are made of scrap balsa pieces that have been glued together. Right: this is the light, simple wing of my Litestik Junior. It uses a full-depth spar, two-piece ribs and turbulator spars to add twist and spanwise bend-resistance. The wing is quite strong and rigid.

If you are building a wing with precut foam-cores, construction is easy once you have learned the skill of applying wing sheeting. However, do not make the mistake of assuming that warps can not be built into foam-core wings. Follow the instructions for building the wing carefully. Use the foam bed and a flat surface to check flatness as you go along. You can remove warps from foam wings with heat and perseverance, but it isn't easy.

Appearance is affected if you sheet the entire wing (built-up or foam-core). Many of us like to see the framing through the covering or even to see the framework itself through transparent covering. The covering adds strength as well as good looks. Choose it well. Some covering materials look great but tend to shatter in even minor crashes. Micafilm by Coverite looks good, adds great strength and only a little weight. It is widely used for the long, thin wings on gliders.

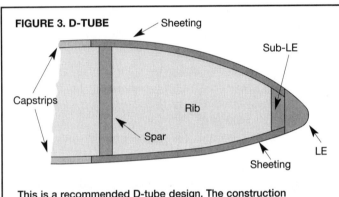

FIGURE 3. D-TUBE

This is a recommended D-tube design. The construction sequence is described in the text. The sub-leading edge adds strength and also makes it easier to install the sheeting.

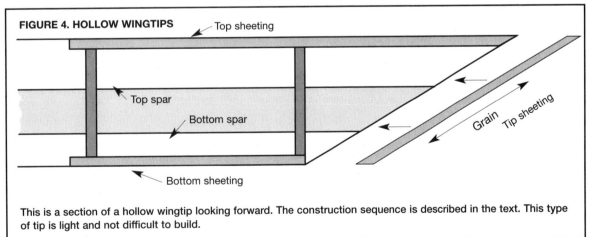

FIGURE 4. HOLLOW WINGTIPS

This is a section of a hollow wingtip looking forward. The construction sequence is described in the text. This type of tip is light and not difficult to build.

CONSTRUCTION TECHNIQUES

It is true that there is little new in this world; most-

WORKSHOP SECRETS 21

 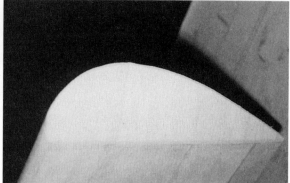

Left: hollow wingtips are strong and not hard to build (see text). Here, the tip bottom has been sanded flat prior to sheeting. Triangular supports can be added, but on small wings, they are not needed. Note the pieced ribs. Once the wing has been covered, only you and I will know it was pieced together of scrap. Right: this is the finished hollow tip. Note that the grain of the sheeting runs spanwise. Repairs are easy: square up the damaged section, fill it with an oversize block/piece of balsa, sand and finish to match.

ly, we combine or rework old ideas. As an exercise, I listed over a dozen, typical, wing-construction techniques. Considering the possible combinations of spars, leading edges (LE) and trailing edges (TE), etc., that are possible, the list could go on and on. The material that follows addresses some of the most commonly used techniques.

• **Spars.** The most practical way to ensure that spars are strong and light is to build them as I-beams. There are many variations on the way model spar beams can be built; some are shown in Figure 1. The strongest I-beams are those with continuous webs. However, many models use interrupted webs: vertical-grain webs are added between the ribs.

The closer the spars come to being full depth, the more they will resist spanwise bending. The simplest approach is a sheet-wood spar that uses the top and bottom sheeting to make it into an I-beam configuration.

Larger wings sometimes use a main spar with a smaller spar between it and the TE. If D-tube wing construction is used, the need for two spars in wings for .60-size and smaller models is questionable. Double spars are useful for fabric-covered full-scale wings because the fabric does not provide strength equivalent to D-tube construction in models.

Ribs are inherently strong chordwise and require little spanwise strength. Capstrips make them stronger in both directions and add to the finished wing's appearance (Figure 1F). There are endless ways to configure ribs; Figure 2 shows four popular variations. Ribs can be made of $\frac{1}{16}$-inch medium balsa for .25- to .60-size wings and $\frac{1}{32}$-inch balsa for smaller wings.

D-tube construction is recommended for all sport/trainer wings. This calls for sheeting the top and bottom from the main spar forward and, for large wings, sheeting at the TE. Shear webbing is required between the top and bottom spars and in front of the TE if the wing is to be D-tubed.

Figure 3 is a cross-section of a typical D-tube wing. The spar and ribs are first assembled on the building board over the bottom capstrips and sheeting. Bevel the sub-LE and install it. Dampen the bottom sheeting and glue it to the rib bottoms and sub-LE. A piece of stripwood under the front edge will help. Dampen the top sheeting and glue it into place. Sand the sheeting flush with the sub-LE. Add the roughly shaped LE, install the top capstrips and TE, and finish the wing as usual. This building sequence allows the use of untrimmed sheeting. Matching sheeting to the LE after it has been installed can be difficult.

Next time you build a wing panel, take the time to check its strength as it is built. It will have to be removed from the building board to do this. With the spars(s), LE and TE in place, try bending the wing spanwise and twisting it chordwise. Add the sheeting and do the bend/twist tests again. And, finally, repeat them when the wing has been covered and finished. If you haven't done these tests before, I guarantee that you will be impressed. A D-tube wing has been described as being as stiff as a 2x10 board.

Incidentally, if you want washout in a D-tube wing panel, make sure you build it in; raise the TE at the tip during construction. More flexible wings can have washout warped in with heat. The amount of washout is not very critical: $\frac{1}{4}$ inch to $\frac{1}{2}$ inch under the TE of the tip of a .40-size model will help prevent tip-stalling. Most important, avoid anything close to washin, for it will cause the wingtips to stall at slow speeds (read "when landing").

• **Wingtips** can easily be carved from balsa using a knife, a block plane and sandpaper (60-grit for shaping and 120-grit or finer for finishing). If you haven't done it, try it; you will be surprised at how good you are at wood carving. Do use soft balsa. If the kit came with hard balsa, save it for another use. Soft balsa will absorb a bump caused by a rough landing. Repairs are easy.

Some of us prefer hollow wingtips. They are lighter (the last place you need extra weight is far from the CG); they "build" easily once you make the decision to go with them. Here's how (see Figure 4): sheet at least the outer tip bay top and bottom, and extend the top sheeting, the LE and the TE to make the wingtip sheeted area long enough to be trimmed roughly to the tip's plan view. Add a piece at the bottom of the outer rib to provide extra gluing surface for the tip sheeting. Sand the tip bottom flat (60-grit sandpaper) using a large sanding block (say, 3x12 inches) or use one of Great Planes' Easy Touch Bar Sanders. The shape of the tip may change somewhat as you sand. This is not important. Once the tip is flat, two or three triangular tip-support ribs can be installed if you want added tip strength. The tip sheeting can be medium to hard with its grain running spanwise. Sand and finish as required. Hollow wingtips are repaired in the same way as solid tips are: square up the damaged area with a knife, glue in an oversize piece of balsa, sand to shape and finish it.

A FAVORITE WING

Not long ago, I lucked in to a Live Wire Champion kit (circa 1979) at a garage sale and decided to build it. I changed the dihedral to about 2 degrees each side, installed trike gear, lowered the engine mount to neatly accommodate an O.S. 25 and modified the wing so I could bolt it on. As predicted, the Champ flies very well. It is another argument for smaller lower-power trainers.

Hal deBolt used my favorite type of wing construction for his Champ—a full-depth spar with two-piece ribs, D-tube LE and hollow wingtips. Even with 4-inch rib spacing, this 54-inch wing is very strong. It is also light, easy to build and looks good. What more can you ask for?

CHAPTER 1

How to strip-plank a fuselage

by John Tanzer

Other modelers often ask how I strip-plank a fuselage with compound curves. Do I have to taper each plank to the rear of the fuselage? No; my method is quite simple: I divide the fuselage into four sections and then glue and pin a strip at the belt line halfway between the top and the bottom sections. Then I glue and pin a strip at the center of the bottom section.

From then on, I simply glue on one strip at a time (one strip up, and one strip down) till they meet at the rear. I then cut the rear of the strips to fit. I do the rest of the sections in the same way, alternating one up, one down, till the section has been filled in.

On my .60-size Zero shown here, I used 3/32x3/8x36-inch-long balsa strips. If some of the strips were too short, I spliced on an additional piece.

I use aliphatic-resin glue because it allows more time to work, and it's easy to sand; just let it dry for 48 hours before you block-sand with 80-grit sandpaper. Do not sand too briskly, as the heat generated can soften the glue and gum up the sandpaper. Fill any low spots (you're bound to have some) with Elmer's light wood filler.

Using this system, I find strip-planking quite enjoyable. The photos show it in detail and are very clear, so why not try it on your next model?

1 To ease the planking process, place the fuselage in a cradle or some kind of padded holder. Here, the left side has been planked, and the right side is ready to be planked. Begin by adding one strip at the belt line and another strip at the bottom center.

BUILDING BASICS

2 For this Zero, I first glued on the ³⁄₃₂-inch sheet balsa saddle for the stabilizer. Note that a ³⁄₃₂x³⁄₈-inch strip has been glued to the fuselage belt line, and another has been glued to the bottom center.

3 As I add strips between the belt line and the bottom strip, their ends come together at the rear of the fuselage. To fit the remaining strips, mark the strips' ends where they overlap the previous strips for the angle cut. When you apply strips, alternate one up and one down.

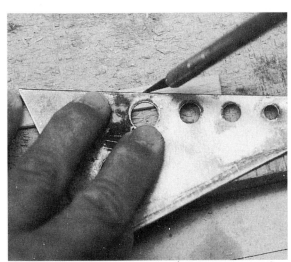

4 Use a straightedge and cut each balsa strip at the mark. Fit each strip into place, and check its fit against the other strips.

5 When the strips are added alternately (one up, one down), the angles on their cut ends create a herringbone effect. The joints that the angled cuts form are very strong and are almost invisible when they've been sanded.

CHAPTER 1

6 The fuselage is now fully planked. Notice that with the joints placed closer to the front of the fuselage, the strips needed are shorter.

7 Notice the difference between the top and the bottom of the fuselage. The top has been block-sanded with 80-grit paper, and the bottom remains unsanded. Aliphatic-resin glue is easy to sand and is perfect for adding planking to fuselages.

8 Here, the Zero fuselage has been sanded and is very smooth. Hardly any filler is required if you make neat, precise joints.

9 These floats for the Navy version of the Japanese Zero "RUFE" were also strip-planked in the same way as the fuselage. Structures of any shape can be planked using this system.

BUILDING BASICS

How to inlay balsa sheeting

by Randy Randolph

For the scratch-builder, trimming wing-center-section ribs to take the top and bottom sheeting is always a hit-or-miss proposition. Even if the ribs are trimmed before they are mounted on the spars, it's difficult to get just the right amount of material removed from the top and bottom of those center ribs. A method suggested by Canadian John Hannah solves the problem simply and easily: don't trim the ribs! The photos show the way.

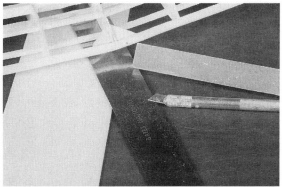

1 Other than the sheet balsa (1/16 inch, in this case), you'll need a metal straightedge to assist in trimming the balsa sheet, a razor knife and a small sanding block (an emery board works very well).

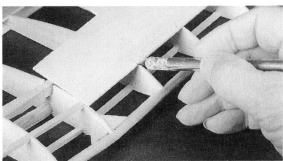

2 Lay the sheet on the area to be sheeted and carefully mark the centers of both ribs on the sheet. A slight cut with the razor knife is an excellent way to mark the sheet. Also, mark the exact width needed (for this piece, the distance between the top spars).

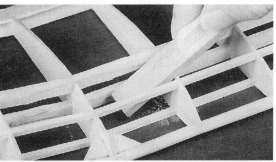

3 Use the sanding block to sand a 45-degree bevel on the insides of both ribs. It isn't absolutely necessary to be exact with the bevel; just try to be as accurate as you can. Sand to the outsides of both ribs; just a few passes are necessary.

4 Now sand the same bevel into the two edges of the sheet. It isn't necessary to sand the sides because they will fit flush with the tips of both spars when the sheet is on the wing.

5 Fit and glue the sheet into place. Additional sanding isn't usually necessary, and the sheet fits perfectly flush with the spars and the tops of both ribs. Aliphatic resin or even Ambroid glue is great for this application and, most of the time, pins aren't needed to hold the piece in place.

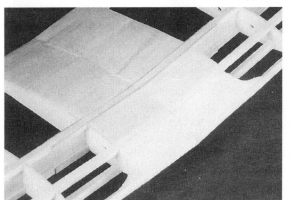

6 The rest of the sheeting is added in just the same way. Actually, the glue joints that are formed between the sheet and the ribs are the same as if the ribs had been trimmed to accept the sheet. This system can be used in just about any area where it is necessary to inlay balsa sheet in a finished structure.

CHAPTER 1

Make built-up truss ribs

by Graeme Mears

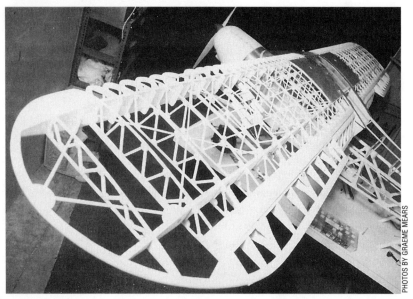

Several years ago, I scratch-built a ⅓-scale Super Cub from Charles Richard Plans. The plans called for built-up ribs in the wings. This was my first scratch-building project, and I was a little unsure of whether or not I wanted to take on all that work. However, Charles had built a number of these Cubs; he had one on landing gear and one permanently on floats. Considering this, I decided to try my hand at building the wings in exactly the way the plans called for. Well, to my surprise, it was not nearly as much work as I had anticipated!

You may ask, "Why do this extra work?" First of all, you have a much stronger rib without adding weight. Then, once you have built all the ribs, the

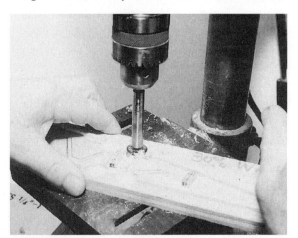

1 Working a jig, the gusset areas are being relieved on a Forstner bit.

wing "builds" very quickly over solid spruce spars. A scale constructed wing is easy to straighten at any stage after construction by means of the cross-wiring. With the cross-wiring system, it was always a problem for me to get the wires through holes drilled in solid ribs; not so with truss ribs.

Preparation is very important if the construction process is to go efficiently. Currently, I am building a 30-percent-scale Waco UPF-7 biplane from my own plans. The only materials used in these ribs are ⅛-inch-square spruce and ¹⁄₆₄-inch ply for gussets.

BUILDING A RIB JIG

I took a photocopy of the wing rib from the plan, and after checking that it was accurate in size (compared with the plan), I stuck it to a piece of ³⁄₃₂-inch aircraft plywood with 3M 77 spray adhesive. Then I used a scroll saw to cut very accurately around the outside rib line. I now have the inside and outside of the rib shape. The female (outer) piece is then glued to a flat piece of ½-inch plywood. The male (inner) part of the rib is then sanded back ⅛ inch to the inside line of the top and bottom capstrips on the power belt sander.

Next, cut out all the areas where the ⅛-inch square truss members will go. Number the parts as you do this because you end up with a jigsaw puzzle. With the aid of pieces of the ⅛-inch spruce as

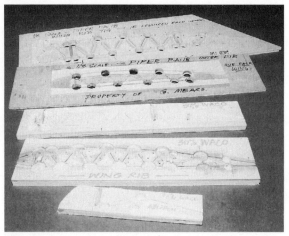

2 Various wing rib jigs for my ⅓-scale Super Cub and 30-percent Waco UPF-7.

BUILDING BASICS

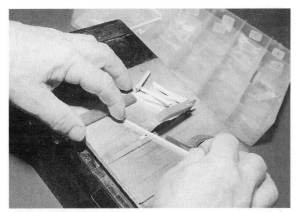

3 Mini table saw has been set to mass-produce wing-rib parts.

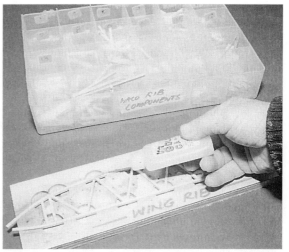

4 Apply thin CA to joints prior to sanding the top face of the rib. Note the compartmentalized box in the background; it holds all the components in order of application.

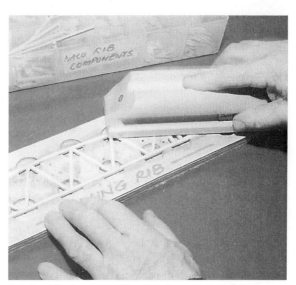

5 Sand the top face of the rib before you install the first set of gussets.

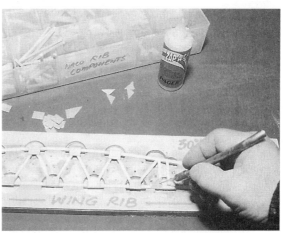

6 Attach the first set of gussets with medium CA.

spacers, glue all the pieces of the puzzle onto the ½-inch plywood base. It's also necessary to have a small section of the front and rear spars glued into place at the correct locations. Now, to prevent the ribs from being glued to the jig as they are built, I relieve all the areas where there will be gussets with a Forstner bit (photo 1). Be careful that you don't take too much of the essential parts of the jig away at this stage. Now the jig is finished (photo 2).

PREPARING FOR CONSTRUCTION

I now make several of the individual components and sort them in a compartment tray. If you analyze the 1/64-inch ply gusset shapes carefully, you will find common dimensions, and you can generally cut long strips of the ply, then snip the pieces to size and shape in a paper guillotine set up with a gauging fence. Remember, these go on both sides of each rib, so make plenty! I set up my mini table saw using the fence and miter guide to accurately mass-produce the 1/8-inch-square spruce truss members (photo 3). I make about 40 at a time. The long, 1/8-inch-square spruce capstrips have to be soaked in water so the sharper curves (at the leading edge) can be made without breaking the strips. Now we are ready to make a rib.

CONSTRUCTING A RIB

It is a very good idea to coat the jig with PVA (mold release) to help prevent the rib from being glued to it. Next, place all the 1/8-inch-square spruce parts in the jig, making sure that everything meets accurately and is flat. Apply a very small drop of thin CA to each joint (photo 4) to hold things together while you are sanding the top rib face (photo 5). With sanding complete, we can glue all the 1/64-inch ply gussets into place with medium CA (photo 6). Be careful to keep the CA off the jig! Carefully place a flat board on top of the rib, separated from it with wax paper, and weigh the rib down. Leave it a few

CHAPTER 1

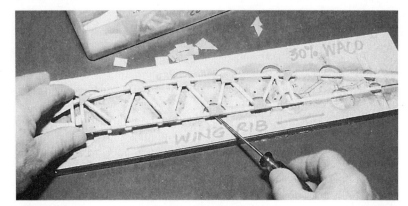

7 Carefully remove the partially finished rib from the jig using a small screwdriver.

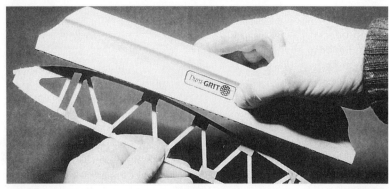

8 Sand the other side of rib with a sanding bar.

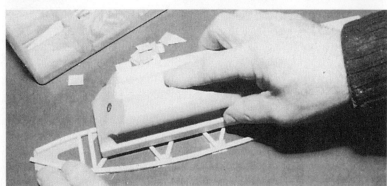

9 Trim and sand to remove all excess gusset material.

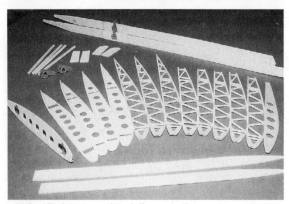

10 All of the completed ribs and other components that make up the bottom wing of the 30-percent Waco.

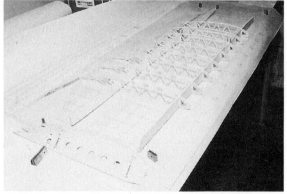

11 The Waco UPF-7 wing during final assembly.

minutes to cure. When the glue has cured, remove the partially finished rib from the jig (photo 7). Now place the rib face down on a flat surface and sand the underside flat (photo 8). When you are satisfied, add the rest of the gussets to the second face. When everything has fully cured, trim and sand the excess gusset material back to the capstrips (photo 9). You have just finished your first scale truss rib! It takes

BUILDING BASICS

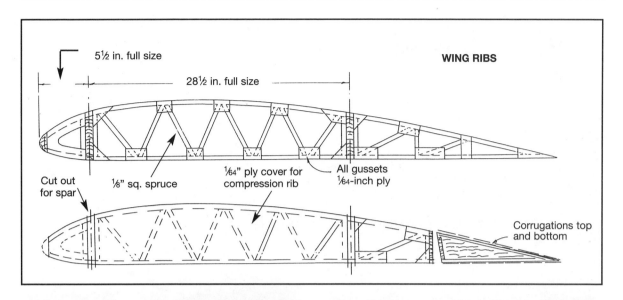

Waco UPF-7 wings (bottom right ready for covering; left partially complete).

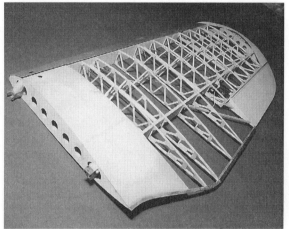

Another shot of bottom right wing (view of top side).

about 20 minutes to make a wing rib in this way. Obviously, if you are building a Stinson Reliant, where all the wing ribs are of different sizes, this method would involve a tremendous amount of work, but for aircraft with a constant-chord wing, it is not as much work as it may appear (photo 10). Some ribs in a scale wing will also be sheathed with a full web of plywood—usually those ribs at the end of a wing panel. Once all the ribs for your wing panel have been made, assemble the ribs and wing spars over the plan and finish building your wing (photo 11). Once it's finished, you may find it hard to force yourself to cover all the beautiful woodwork. For a scale model builder like me, who is a bit over the edge, there is a great deal of satisfaction in knowing that, underneath, the surface is as scale as it can be.

CHAPTER 1

Sanding basics
by Mariano Alfafara

The author spent 18 months building and sanding this P-51.

It has been said that you can measure a builder's skill level—without seeing a model—by examining the sanding tools in his workshop. I have visited many workshops, and I agree that the best builders/finishers use only premium-quality sandpaper and have made a wide assortment of specially designed sanding tools. Of all the skills that we modelers must master, sanding is probably the most important.

Before we can develop good sanding techniques, however, we must be able to select the correct materials for the job. How do you choose the correct sandpaper for a given task? Novice modelers may find the many choices somewhat confusing: production paper, garnet paper, or flint paper? Silicon carbide, emery, or aluminum oxide? Open coat, closed coat, or no-load? What size grit or grade? Wet or dry? What is A, C and D paper?

A little baffling? Possibly, but only if you can't identify the various sandpaper types and designs. Therefore, our objective is to simplify this selection process by learning about abrasives and their designated applications. The next time you're at the hobby shop and need sandpaper, you'll be able to say something like, "Give me a couple of sheets of open-coat, aluminum oxide, 320-grit, on A paper."

The sandpaper that we model builders use is technically called "coated abrasive sheets." To simplify this discussion, we will continue to use the common misnomer: sandpaper. Three basic components make up sandpaper: an abrasive, a bonding agent and a backing (see Figure 1).

To select the correct sandpaper for a given task, we only need to understand the information on the back of the abrasive sheet.

ABRASIVE MATERIALS

• Natural minerals (garnet, crocus and emery). Garnet is a widely used mineral of medium hardness; it has a sharp form and good cutting edges. It's a good choice to use with balsa and other soft woods. It is paper-mounted, orange-brown, economical and readily available. Crocus and emery are cloth-backed abrasives that are designed for metal finishing. Other than polishing the

This photo reveals that your hand makes an unsuitable sanding block. The black silicon-carbide paper reveals the uneven removal of material, which results in an uneven surface on the model. A skilled builder uses specialty sanding tools to remove material more uniformly.

BUILDING BASICS

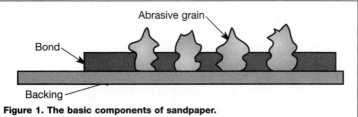

Figure 1. The basic components of sandpaper.

occasional small metal part, there is limited use for these products in our workshops.

• **Synthetic minerals.** Aluminum oxide and silicon carbide are generally the only two synthetic minerals that suit our modeling applications.

Aluminum oxide is gray-brown, or sometimes gold. It is extremely hard and resistant to wear. 3M markets this abrasive under the trade name of Three-M-ite (not to be confused with their Tri-M-ite). Norton markets its aluminum oxide under the trade name Adalox. I've used both of these paper-backed products on hardwoods, resins and paints with equally exceptional results. Production paper is made with aluminum oxide.

Silicon carbide is the hardest and sharpest of the minerals commonly used in coated abrasives. It is brittle and fractures into sliver-like wedges as it works. You could say that it sharpens itself as you use it—pretty cool, huh? It's ideal for both ferrous and non-ferrous metals, e.g., steel, aluminum and brass, and also for plastic, wood, resin, rubber and both hard and soft paints. Silicon carbide is superior to other abrasives in its capability to penetrate and cut faster under light pressure. This characteristic alone makes it very well-suited to modeling. The 3M trade name for silicon carbide is Tri-M-ite; the Norton trade name is Durite. Both brands are gray. Although it is more expensive, silicon carbide is certainly the abrasive of choice in my shop.

An additional note on types of abrasives: remember flint paper, that sandpaper we used in shop class at school? Well, if you have any of that stuff in your shop, throw it away or give it to someone who does craft projects.

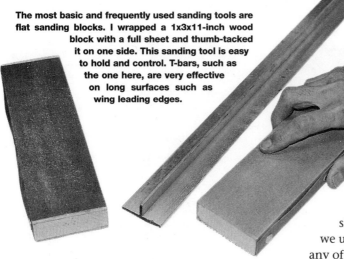

The most basic and frequently used sanding tools are flat sanding blocks. I wrapped a 1x3x11-inch wood block with a full sheet and thumb-tacked it on one side. This sanding tool is easy to hold and control. T-bars, such as the one here, are very effective on long surfaces such as wing leading edges.

COAT TYPES

Abrasives are coated onto a backing by two methods: open coat and closed coat. Open coat indicates that the abrasive grains cover 40 to 70 percent of the backing. If the material being removed has a tendency to load or clog, an open-coat sandpaper is the proper choice. Loading is the tendency of the material being removed to build up on the abrasive grain. We should use open-coat sandpaper for most of our modeling needs. Closed coat indicates that the abrasive grains cover 100 percent of the backing. A closed

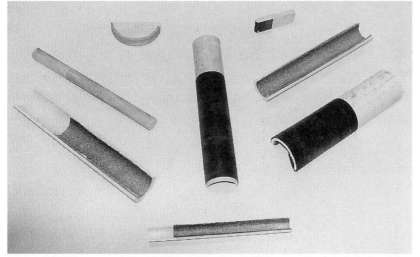

Every aircraft has a number of inside and outside radii that need to be shaped and finished (fillets, wingtips, leading edges, etc.). Here are some radii sanding and shaping tools that I made from dowels, plastic pipe cut lengthwise, or whatever else I could find to match the specific radius I was forming.

WORKSHOP SECRETS **33**

CHAPTER 1

coat is the proper choice when loading is not a problem, such as in metal finishing.

WET SANDING

For wet sanding, a waterproof bonding agent is used with silicon carbide. The 3M trade name is Wet-or-Dry Tri-M-ite; Norton calls it Tufbak Durite. Both are black. As their names imply, they can be used in either wet or dry sanding applications. These products really excel at wet sanding, however. With the non-clogging action provided by water and the superior cutting ability of silicon carbide, these papers can quickly remove a lot of material.

The paper mask protects the rear portion of the leading-edge sheeting, while the sanding block levels the uneven rib capstrip.

An aerosol-applied adhesive is ideal for attaching sandpaper to wood or plastic tools. Note the sandpaper glued to the bottom portion of the can; its diameter was an exact match to the diameter of an open cockpit on a Ryan STA that I was building. The can was the perfect tool to sand the edges of the cockpit opening.

The makers of Hobby Poxy paint recommend that you do not dry-sand their paints with wet/dry paper. They feel that the waterproofing agents in the paper could get on your finish and interfere with paint adhesion. I have used this sandpaper both wet and dry and have not had any problems, but caution is advised. In any case, you should always wipe the model surface with alcohol or thinner after sanding to remove any residue.

BACKINGS AND BONDS

Backings are the platforms that carry and support the abrasive mineral grains. There are five categories, but we are concerned only with paper, and occasionally cloth-mounted emery. Paper backings are classified by their weight as A, C, D, E and F.
- **A paper**. The lightest and most flexible paper, coated with fine mineral grains.
- **C and D paper**. Intermediate-weight paper, coated with medium mineral grains.
- **E and F paper**. The heaviest papers; can be coated with almost all grit sizes.

Sponge or foam sanding pads are used only after your surface has been sanded with sanding tools. Sanding pads should be used only to remove glaze and minor imperfections on relatively smooth surfaces.

I want to use sandpaper with a backing that will conform to curves and radii, so in most instances, I use A-weight paper. I use C- or D-weight sandpaper on straight, flat sanding fixtures such as sanding blocks and T-bars. (By the way, there isn't a B weight; I don't know why.)

BUILDING BASICS

Bonds are two or more layers of adhesives that anchor the abrasive mineral to the backing. There are many types, including combinations of glues and resins. Bonds are a significant factor in the performance of the abrasive. As modelers, however, we select a sandpaper by its abrasive mineral, grain and backing, not by the bond. Therefore, you need only know that for wet-sanding you will need a waterproof bond such as 3M Wet-or-Dry, Tri-Mite or Norton Tufbak Durite.

■ Sandpaper Selection

When you select sandpaper, remember the following factors:

- **Material to be removed.**
  Hard or soft wood, resins, plastic, paints? Will the material load or clog the abrasive? The answers will determine which abrasive and type of coat you'll need.
- **Shape of working surface.**
  Flat, compound curve, or radius? These factors will determine a backing selection.
- **How much material is to be removed; what kind of final finish is desired?** This will determine grade or grit selection.
- **Will you wet-sand? If so, ...**
  you will need a waterproof bond.

Abrasive (color)	Applications
Garnet paper (orange/brown)	All types of soft wood
Aluminum oxide (gold/brown)	Wood, resins, sealers, paints, primers
Silicon carbide (gray)	Wood, resins, sealers, paints, primers
Emery (black)	Ferrous and non-ferrous metals

Suggested applications for wet-sanding

Silicon carbide (black) (3M Wet-or-Dry, Tufbak Durite)	Lacquer, paints, resins

Selection of paper-backing weights

A-weight	Compound curves and radii
C- and D-weights	Flat work

Selection of grades (grit)

50, 60 and 80 grade	Heavy stock removal
100, 120 and 150 grade	Medium stock removal
180, 220 and 240 grade	Light stock removal
280, 320, 360, 400 and 66 grade	Finishing

GRADES

The grade size refers to the size of the abrasive mineral grain. This size is determined by the number of grains that will fall through a designated number of openings in a screen mesh per square inch. There are 22 grit sizes, or grades, as they are commonly called. They range from the coarsest (12 grade) to an extremely fine powdered form (1,500 grade). Modelers usually use grades of 60 to 80 for rough stock removal and up to around 400 to 600 for those super finishes that you see in the winners' circle at major contests. Grades finer than 600 are rarely used in our hobby; nevertheless, my friend George Maiorana used 1,500-grade wet or dry sandpaper on the clear acrylic windows and gun blisters in his Top Gun B-29. There are other exotic grades designed for very special applications; I once read that NASA uses a 50,000 grade.

SUMMARY

Whether you're preparing an airframe for paint or iron-on covering, nothing can hide or cover up a poorly sanded surface. Many aspiring builders often underestimate the significance of this fundamental skill. When I asked Darrell Rohrbeck to summarize his experience building his magnificently finished ¼-scale Long EZ, which won first place at Toledo '97, he recalled that the project was "... about 10 percent building and 90 percent sanding."

You should have a wide assortment of sanding tools in your shop. Here are some of the tools that I made and used to build my all-wood P-51.

CHAPTER 1

Sheet foam wings with plain brown paper
by Bertil Klintbom

You can create a perfect finish with brown wrapping paper. This method is easy, low-cost and creates a clean, strong surface that's very tough and resists punctures. If you like, you can cover an entire model this way! I covered the wing and parts of the fuselage of my 1/8-scale Casa 212 Aviocar this way.

1 Add leading and trailing edges and the wingtip to the foam-core. Mix regular white glue 50:50 with water and add some food coloring. Cut the brown wrapping paper so that the top sheet is 1 inch wider than the foam-core, and the bottom sheet is slightly smaller than the chord of the foam-core. If there is a "grain" to the paper, it should be positioned spanwise. Apply the glue mixture to the foam-core; you can see why you need to color it! Coat both sides of the foam-core and the matte side of the paper with the glue mixture.

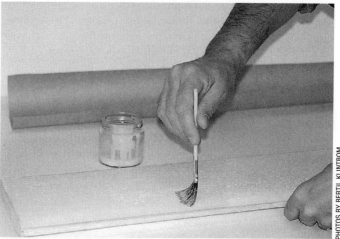

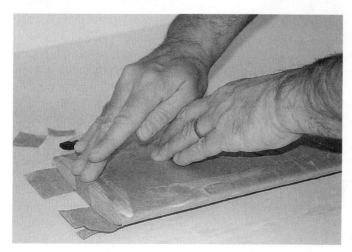

2 Position the paper on the foam-core. Start with the underside and continue with the upper surface. The upper sheet should overlap the edges and the bottom sheet. Smooth out the paper with your hands and work quickly, before the paper gets too wet. Make sure to cover both sides of the foam-core! Cut the paper at the wingtip and root and fold small pieces of it up over the side to the bottom surface. If you have a rounded wingtip, simply cut darts in the paper and let the pieces overlap.

BUILDING BASICS

3 After you've covered the wing, it will start to wrinkle and will look awful: a total disappointment! No problem; hang it to dry for 24 hours, and it will become smooth again.

4 After the wing has completely dried, coat it with a mixture of 50:50 water and white glue. The cover will wrinkle again, but hang it to dry, then repeat this process once more.

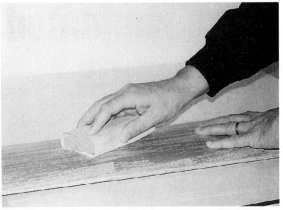

5 After the wing has completely dried for the third time, dry-sand it. Be careful not to cut into the paper surface.

6 Fill out the paper joints with lightweight filler, and after the wing has dried, sand the joints and add more filler if needed.

7 When you are satisfied with the surface, continue with a primer and paint as usual. If you wet-sand the wing, be sure you have a good coat of primer on the surface.

WORKSHOP SECRETS **37**

CHAPTER 1

Make laminated paper parts

by Harry B. Cordes

Here's a technique worth considering the next time you need a special shape for a cowl, nacelle, radome, turtle deck, or fairing. The result is light, easy to work with and costs practically nothing! All you need is newspaper, Titebond II glue and a mold form shaped out of scrap wood. The process is similar to fiberglassing, but you use newspaper and glue instead of glass cloth and epoxy. It is also a useful technique for producing complex shapes that can't easily be made by vacuum-forming.

The cowl and the nacelles on my PuddleMaster have distinctive shapes; forming each requires a different approach, especially in the layup process. The following description shows you how to lay up most shapes. You may need to experiment when you cut the paper laminations, but newsprint is cheap.

1 First, make the layup forms. I made the nacelle form ¼ inch larger in diameter than my motor's outside diameter (o.d.) to allow cooling air to flow around it. To allow the finished layups to slide off, I tapered the front of the nacelle form about 3 degrees, then glued it to a wooden wing form and faired it into the wing using wood putty. I made the cowl form ³⁄₃₂ inch larger in diameter than the nacelle to allow for the build-up in the nacelle's wall thickness; I also tapered it so the finished cowl would be removable. The ⅛-inch-diameter dowel in the front of the cowl form holds the paper laminations in place during the layup process; it also centers a guide washer for cutting out the center of the finished cowl. To seal the grain, I gave the forms two coats of shellac, fine-sanded them and finished them with a heavy coat of paste wax.

2 Next, I applied a layer of wax paper around the cowl and the front of the nacelle—held in place with a dab of paste wax—as added insurance that the layups would be easy to release from the forms.

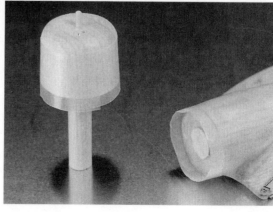

BUILDING BASICS

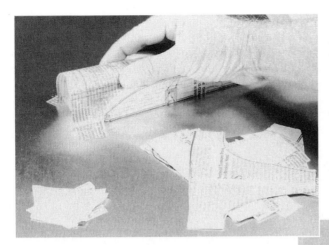

3 Before laying up the nacelle, I experimented with several different paper shapes to find the best fit for the contour. The two shapes shown worked nicely. The larger one wrapped completely around the nacelle; when it was wetted, I was able to work it into the wing/nacelle fillet with minimum tearing. (Some tearing is tolerable as long as the tears in one layer do not overlap those in the next. Reinforce torn areas with interleaved paper patches or strips.) The smaller shape covers the lower side where the nacelle is faired over the leading edge of the wing. Because the finished part is trimmed to size after the layup has cured, I cut the shapes oversize.

4 To laminate the nacelle form, I soaked a layer of paper in water and smoothed it into place, covering the form on the top and bottom. Next, I liberally applied a 50/50 mix of Titebond II glue and water to the still-damp surface of the first layer. Then, I put the second layer into place to dry. The fresh glue below permeated the second layer and made it more pliable as it was smoothed on.

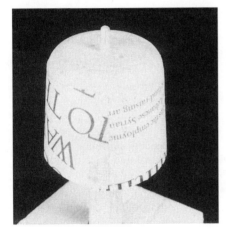

5 A round plastic rod or an artist's brush handle makes a good tool to work into fillet areas and to burnish wrinkles. With another layer of glue mixture on the second layer, I applied the third layer (again, dry). Continue this process until you reach the final layup thickness. I used six layers on the nacelle to get a buildup of about 0.04 inch. After adding each layer, work it tightly to the form and to the other layers so that air won't be trapped between them; burnish wrinkles as much as possible. Make sure all overlaps are covered with glue, and stagger overlaps to obtain a reasonably uniform thickness. Reinforce tears with patches as required. After you've smoothed the final layer into place, apply a final coating of glue, and set the work aside to dry and cure.

6 To start the cowl layup, I wrapped a single wetted strip around the straight portion of the cowl form and held it in place with glue at the overlap.

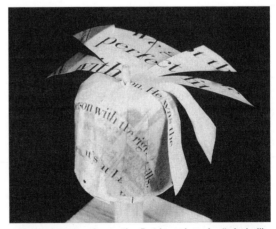

7 Apply more glue to the first layer in a dry "windmill-blade" pattern as shown. Lay down the "blades" one at a time; use glue to bond each to its neighbor. To hold the pattern in place, a cross-slit in the pattern's center slides over the 1/8-inch-diameter dowel.

WORKSHOP SECRETS **39**

8 Next, apply more glue to this layer; then tightly apply a wraparound layer, with axial flaps cut at the front. Lay down the flaps around the front curve of the cowl, using a liberal amount of glue. Lay up the windmill and wraparound patterns alternately until you achieve the desired buildup. As with the nacelle, smooth each layer tightly, burnish wrinkles, and apply a final coat of glue to the last layer.

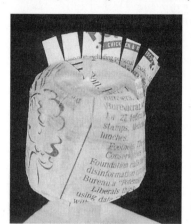

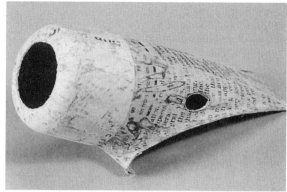

11 After I had epoxied the cowl to the nacelle, I applied several coats of automotive filling primer; between coats, I dry-sanded with 600-grit paper, then I used acrylic enamel as the final coat. I then epoxied the aluminum-tube stacks that exhaust the cooling air into the side openings and waterproofed the interior with epoxy paint.

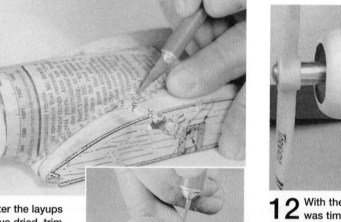

9 After the layups have dried, trim them. A no. 11 hobby knife works well for trimming edges and cutting openings.

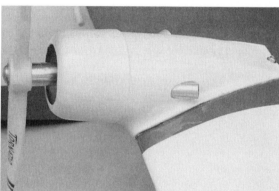

12 With the finished cowl and nacelles installed, it was time to fly! The light, strong cowl and nacelles add much to the model's appearance. Try this laminating method the next time you need to make a cowl or fairing. Your friends won't believe you made it out of the morning newspaper.

10 After I trimmed them, I sanded the layups smooth with 200-grit garnet paper to remove bumps and wrinkles; next, I removed the layups from their forms. Working a knife blade around the edges helps to loosen the layups; with some gentle twisting and pushing, they will pop off. Two 1/8-inch-square spacer strips were glued inside the nacelle at 45 degrees from vertical to align it with the motor o.d. These spacer blocks, together with the faired wing overlap at the bottom, hold the nacelle firmly to the wing; only a single screw at the back is needed to secure it.

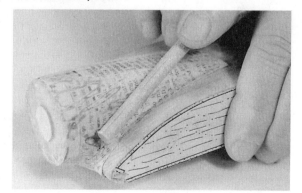

BUILDING BASICS

Mount wings using differential screws
by Al Ehrenfels

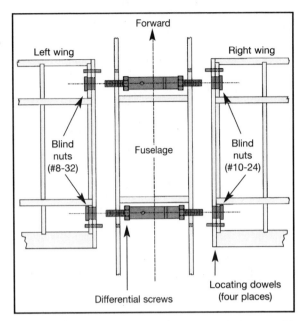

The wing-mounting method shown on the plans of my ¼-scale Cub left me cold. After considering turnbuckles and finding only some rather crude and bulky examples on the market, I developed an alternate approach that works surprisingly well and may just be what you need for your next project.

The method I use employs a differential screw in place of a turnbuckle. The differential screw is an old engineering concept, and this is merely a new application. Before you snort "So what!" and move on, consider this: it requires no special taps; it uses readily available material; it needs no fancy machining; and it can develop an incredible amount of retaining force.

The accompanying sketch shows an aluminum rod with lengths of threaded rod fitted to each end. Different threads are used on each end, hence the name "differential screw." It can be any combination of threads, providing they are different on

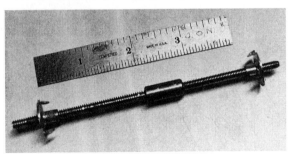

each end of the assembly. In my application, 8-32 and 10-24 threads were used, since that's the threaded rod the local hardware store had on hand. The matching blind nuts and the aluminum rod were available at the same place. The blind nuts are sold as "T-nuts" in the local hardware store, so try that moniker if "blind nuts" draws blank stares.

Looking at the drawing of the installation in my Cub, you will see that I used two screws passing through, but floating in, the fuselage. Blind nuts are fitted to the root ribs of the wing, and locater dowels in the root ribs fit matching holes in the fuselage.

To assemble the plane, I hold the left wing against the side of the fuselage and turn the 8-32 screws into the blind nuts. I screw them in until they are almost flush with the fuselage on the right side. Then I hold the right wing against the fuselage and proceed to turn the screws into the 10-24 blind nuts on the right wing. You might think this would result in my unscrewing the left wing at the same time, but note that the screw is moving into the right wing faster than it is moving out of the left one. That's the secret to this approach.

The reason it works so well is clear if we do a little arithmetic. An 8-32 thread has a pitch (advance per turn) of 0.03125 inch, and a 10-24 thread has a pitch of 0.04166 inch. Thus, as I am screwing into the right wing, both wings are being pulled together by 0.0104 inch (0.04166 minus 0.03125 inch) on every turn of the screw, and that's the pitch of a 96-threads-per-inch (tpi) screw. A few turns of the "turnbuckle," and the wings are pulled to the fuselage with the mechanical advantage of a 96tpi screw, and that amounts to about 100 pounds, using only fingertip pressure to tighten the screws.

For the occasions when my hands were slippery with oil, I drilled two cross-holes in the rod so that I could use a short piece of music wire to turn the screws. Go easy, or you'll have a crushed fuselage.

Does it work? My 9-foot Cub can be picked up by the wingtips, struts removed, with virtually no sag. When I recently walked away from a skirmish with a runway marker light (I swear the thing was magnetic!) with only minor cuts and bruises to the Cub, I knew for sure that I had a good thing going with this design.

I recommend it for a simple, but very effective, solution to the problem of attaching wings to your aircraft's fuselage.

CHAPTER 1

Work with EPP foam
by David M. Sanders

In the world of model aircraft construction materials, there's a new kid on the block: expanded polypropylene (EPP). The latest generation of slope combat sailplanes and sailplane trainers is made of this material, which has proven to be a great advance in the search for the truly indestructible airplane. Want to know more? Check this out ....

ITS PROPERTIES
Unlike the white, beaded foams we've known for years, which are expanded polystyrene (aka EPS or Styrofoam), EPP is made of polypropylene, a very long-chain polymer molecule. As a result, EPP is not crumbly like EPS and is much more difficult to break or tear. It also has incredible memory; a piece squashed between your fingers will spring back to its original form in a matter of seconds! EPP looks a lot like regular beaded EPS and is made by a similar process in which beads of the plastic material are

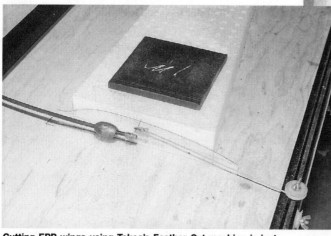

Cutting EPP wings using Tekoa's Feather Cut machine is just like cutting polystyrene foams, except the wire must run much hotter to get reasonable speeds and good clean-out of the kerf.

steam-injection molded in a large die to form billets (known as "buns" in the industry). And here's the real topper: it's light! EPP comes in two densities—1.3 lb./sq. ft. and 1.9 lb./sq. ft.—virtually the same densities we're used to working with already.

OK, so what's the downside? As with anything else, there's no free lunch here. EPP can be difficult to work with. Many common adhesives—including our cherished CA—are useless on EPP. Epoxy is of limited value, too. EPP foam is impervious to practi-

cally any solvent or corrosive agent—even battery acid! In addition, you'll need significantly more heat for hot-wire cuts (this, however, has hidden fringe benefits I'll discuss later). Finally, the molded billets are almost always crooked, and their dimen-

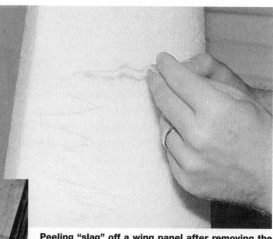

Peeling "slag" off a wing panel after removing the saddle. By carefully setting the cutting wire's temperature, you can have the slag bunch up into easily removable strands. The part's surface will be as flawless as your templates—even on steep tapers.

sions vary wildly; you must cut straight parts out of misshapen blocks.

CUTTING IT
I cut EPP using two methods: hot-wire cutting (in a well-ventilated area) and band sawing. I consider these the safest methods. Pat Bowman, maker of the Ruffneck and owner of Bowman's Hobbies, told me he tried exactly once to cut EPP with a table saw, and he recommended that I not try doing that myself! Lex Liberato of Studio B, maker of the Foaminator, uses a router in his EPP armament, but I tried using a router once and consider it suicidal (just my opinion).

Hot-wire cutting takes massive amounts of power. I run .025-inch, stainless-steel fishing leader on my bows, and my 36-incher needs about 50 volts at 10 amps to get hot enough to cut EPP efficiently. (Note: if you accidentally grab

BUILDING BASICS

the bow at the energized ends, you'll "ride the lightning" in a big way!)

After a cut, the parts don't come apart cleanly; they sort of stick together. When you peel the parts away from each other, little hairs of melted foam remain on the part, the bed, or both. I call this stuff

Spar slotting with a hand-held bow. The simple jig is easy to construct and very precise. Hot wire is my favorite method of cutting slots, but routers and soldering irons can also be used to good effect. Be careful using power tools; EPP likes to grab bits and blades.

"slag." If the cut leaves slag all over the part, then your hot wire is running too cold. If you run it hotter, you can get the slag to bunch up into one or two big strands of melted material that peel away fairly easily. If you run it too hot, though, the slag strand will take some beads of foam with it when you peel it off; this can ruin the part. The best way to find the right temperature setting is by experimenting, but if you take about 15 seconds per inch of cut, you're in the ballpark. Also, you have to pay particular attention to your templates because the high heat settings can actually melt a groove in the guide edge if the wire stalls. Be vigilant always! I use countertop-grade Formica for templates.

BAND SAWING IT

Band sawing this stuff is a breeze and is actually my favorite method of cutting EPP. (A scroll saw also makes short work of it.) I use very coarse blades that have four to six teeth per inch; these seem to clean out the kerf most efficiently and allow extremely tight radius cuts. One thing to watch for while cutting on the band saw is scrap pieces that can be sucked through the lower guides and then melt from the blade friction. This can leave a sludge of EPP on the blade and wheels, which can create extra tension and cause your blade to break. Although not a huge danger, it's certainly an inconvenience (I go through a lot of band saw blades). Although EPP has a tendency to dull cutting edges,

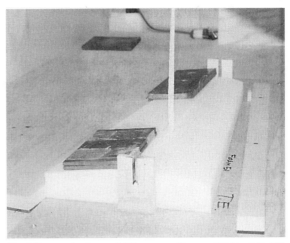

Peel the waste from slotting operations quickly or it will tenaciously stick in the slot as the hot foam cools. Notice the waste strip's incredible flexibility; I can easily pull it out in one, long piece.

I've found that even aged band-saw blades can still make clean cuts in the foam.

SHAPING AND SANDING IT

Yes, you can actually sand this stuff, but it sure doesn't sand like wood! Very gritty (80 to 150) paper can cut pretty quickly. My favorite way to do bulk shaping is with sharp (I said sharp!) razor knife blades. Keep handy either a healthy supply of blades or a sharpener. Never struggle with cutting; if the blade hangs and pulls off chunks, it's dull. Sharpen or replace it. Another tool that I've heard works well is a good ol' Stanley Sureform rasp used with light pressure. You could also use a miniature hot-wire cutter.

The best way to sand EPP is with power equip-

Band sawing an EPP fuselage is joyfully easy. Even a scroll saw can cut the foam very cleanly and effectively. Coarse-tooth blades work best to ensure good clean-out of the kerfs and minimal binding of the blade in its guides.

CHAPTER 1

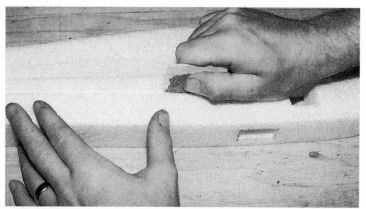

Using gritty sandpaper wrapped around a piece of wood makes it easy to cut slots. Note cavities in fuselage for installation of radio equipment. Neatness pays!

ment. A benchtop disk or belt sander will walk right through a piece of EPP and leave a very clean surface that's nearly as clean as a saw cut. One problem with power sanding, though, is a tendency for the foam to "grab" the wheel or belt. This can also ruin a part and scare the hell out of you when the machine pitches the part across your shop. Always use a slow feed rate and hold the part firmly.

You'll also be cutting cavities to install radio equipment and stuff. Good tools for this are a sharp razor knife or even a pen-type soldering iron. Be careful with the soldering iron because the foam

To ensure a good fit, it's helpful to lay out your radio gear before you cut the cavities. Spacing of ⅜ to ½ inch between cavities is usually sufficient, but more is better.

retains heat for a long time and can burn you.

REINFORCING IT

Some EPP foam components—especially wings—will require stiffening. This is generally accomplished by putting sticks of wood or carbon fiber inside the foam or adding strapping or vinyl tape to the outside.

Other than pushrod housings, most conventional EPP combat planes don't have any rigid stiffening agents in the fuselage construction. Tape does the job of keeping the fuselage stiff and break-resistant. Wings, on the other hand, do require a spar and are often equipped with wooden ailerons or trailing-edge strips like that of traditionally constructed aircraft. I prefer using basswood spars. A benefit of using rectangular materials like wood is that the foam can't rotate around the spar as it can if you use a tubular spar. This, however, can also largely be controlled by using tape. Other makers use spruce or even carbon-fiber spars.

Three types of tape are commonly used now. The first is vinyl, which is used on Trick R/C's Zagi and Razor combat wings. Vinyl imparts plenty of torsional strength for lightly loaded airplanes. Some conventional models rely on vinyl tape, too.

The second type of tape that's often used is strapping tape, the fiberglass-reinforced packaging tape sold by stationery or office supply stores. I don't skimp on tape; I look for high-quality stuff with large section filaments. The brands with fine, closely spaced filaments just aren't strong enough.

Finally, there's plain, unreinforced packing tape. Mark Mech of Aerofoam, maker of the EPP foam F-4U Corsair, likes to cover his airplanes entirely with Mylar packing tape and then uses sign vinyl as a final skin.

One note about the packaging tapes: they disintegrate quickly in UV radiation, so if you want a long-lasting airframe, you need to cover the tape with something. This leads us to final finishing materials....

COVERING IT

There are quite a few options other than the vinyl tape mentioned above. All require that a coat of contact cement—either Weldwood or 3M Super 77—be applied to the raw foam before covering; they'll stick to the Mylar tape pretty well on their own. Also, be sure not to run your iron cold! EPP is not nearly as sensitive to the heat of the covering iron as Styrofoam. I run my iron at exactly the same temperature that I use to cover wood, which is pretty hot. I set my Top Flite iron at 2.75 on a scale of 0 to 3. One final note on coverings: I like materials I can shrink over and over

44 WORKSHOP SECRETS

Building Basics

Sticking it Together

A big challenge in developing EPP airframes was figuring out how to join parts. Here's a rundown of adhesives that work for various applications in the airframe:

My favorite sticky stuff for EPP foam. All are available at hardware stores.

- **Household Goop.** This can be found in most hardware and craft stores and comes in a bright purple tube. It's probably the best all-around adhesive for joints that can't be reinforced in any other way, such as with tape or covering film. I use Goop for attaching wings and tail feathers to my planes. It comes in many varieties, but I think they're all pretty much the same. They all have the unmistakable smell of Ambroid cement and are based on the same solvents.

- **Shoe Goo II.** Similar to Goop, but a lot more viscous and a little tougher to work with. Both Goop and Shoe Goo have very short shelf lives, but Goop seems to last a little longer.

- **Hot-glue gun.** This is by far the most popular adhesive because it's fast. Goop and Goo can take days to cure, but hot glue only takes a minute or two. If you work fast, even large components like spars and trailing edges can be installed with hot glue. I use it for gluing into the airframe anything that will later be taped or covered over, such as spars and servos.

- **Liquid contact cement.** Products such as Weldwood general-purpose contact cement are excellent for bonds that require no prepositioning. It's also good for spreading over the foam before covering.

- **3M Super 77 Spray Adhesive.** I use this as an intermediate adhesive under tape or covering film. With the film adhesive and the heat of the covering iron, the Super 77 provides a very good bond on the foam.

- **Heat.** This is seldom used but is effective for direct foam-to-foam bonds.

- **Carpet tape.** This is surprisingly strong and can be used to attach wings and tail parts to fuselages. In addition, it's easy to replace, if necessary. To use carpet tape effectively, the joints have to fit together really well.

You'll make two kinds of joints on today's modern combat planes: foam-to-foam and tape-to-tape. The tape used to reinforce airframe parts has a Mylar matrix and is as difficult to bond as the foam, so I prefer Goop for tape-to-tape joints. Goop is also effective for installing tail parts to fuselages and winglets to wings.

again. If your plane gets beat up and wrinkled after a day's hard use, a quick once-over with the iron can get it looking as good as new.

- **Ultracote original recipe (not "Plus").** My personal favorite for a scale finish. I've found it gives me good results every time, has predictable characteristics and lasts a long time. A drawback is that it's one of the heavier materials.

- **MonoKote.** Useless on EPP. Don't even try it; it just won't stick.

- **21st Century Film.** Works very well but is cantankerous about heat settings. If you're already a 21st Century user, you'll be in heaven.

- **Model Research Labs 1.5 mil Mylar.** This stuff is awesome; it's very light and incredibly strong. It goes on just like any other covering film. A big drawback is that it only comes in clear.

- **Micafilm.** Yes, Micafilm! It works quite well over Balsarite or 3M 77 Spray adhesive. It can be tricky to work with, has a limited semitransparent color selection and low shrinkability, but it's very lightweight. A good possible choice for foamie HLGs.

- **Solartex.** This works well but is pretty heavy. It's great for painting over. One drawback is that after it has been painted, reshrinking the film can be a real bear because the heat ruins most paints.

- **Sign vinyl.** Comes in many weights and colors. Can be applied over Mylar packing tape. Looks good and is sun-tolerant, but it's definitely on the heavy side.

CHAPTER 1

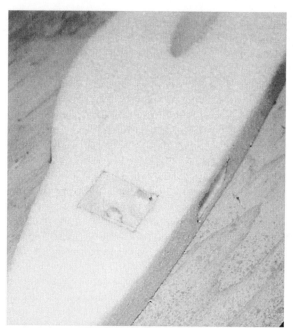

These radio cavities have been cut with a sharp razor knife. The holes between the cavities were made with a sharpened brass tube.

One trick to covering with iron-on films is not to shrink a part until it's completely enveloped in the film. If you try to shrink the pieces one at a time, they'll just keep shrinking until they reach the lower limits of their dimensions and they'll never get tight. Instead, cover a part completely, with only minimal shrinking in the middle of each piece of covering. After all the edges of the separate pieces of covering are sealed to each other, shrink the whole part at once. This can result in incredibly tight, smooth covering jobs.

HOW DO I GET THIS STUFF?

Kit manufacturers and hobby suppliers of EPP get the foam from packaging manufacturers. The catch is that they buy it in huge quantities. I buy this stuff by the truckload to make my kits, as do other makers and suppliers. Don't despair; you can get just a couple of pieces at a time from Aerospace Composite Products, Superior Aircraft Materials and Dave's Aircraft Works (my company). Call for prices and availability. Also, EPP costs about twice as much as ordinary EPS, so get ready for a mild case of sticker shock.

This should give you an idea of what to look forward to in an EPP foam project. It's definitely different and requires the development of new skills, but after you fly an EPP airframe and dork it a few times, you may never want to go back to wood and fiberglass! Good luck and happier landings.

BUILDING BASICS

Working with metal
by George Wilson Jr.

Working with metal is not difficult, even for modelers who have limited workshops. Much can be done using just a few hand tools. Scratch-builders can build with "store-bought" wheels, landing gear, linkages, etc., or be "purists" and make all or almost all that is needed. It's best to start by buying and shift to making as you gain experience; to be a scratch-builder, you don't need to take a trip to the South American jungle with an ax in hand!

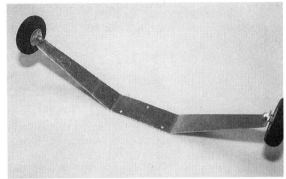
A landing gear cut and bent out of hardened aluminum. Drill the holes while the blank is still flat.

This sheet-metal bending brake makes very nice bends because you can control the bend radii.

WORKING WITH SHEET METAL

Aluminum is the sheet metal used most in model airplanes. Soft aluminum can be bent and cut easily. Hardened varieties resist being bent and are more difficult to cut. Both can be cut with a hacksaw by hand or by using bench, jig, or band saws. Rough edges can be smoothed using relatively coarse files or sandpaper.

To prevent "stress cracking" and the consequent breakage at the bend line, avoid making sharp bends. The harder the metal, the more subject it is to stress cracking. On the other hand, hardened aluminum is "springy" and therefore makes good landing-gear struts.

If you are fortunate enough to have a "metal-bending brake," you can set up the bend radius before making the bend. If you use a bench vise to do your bending, be sure it has a round jaw. You can make a round jaw by filing the corner of a piece of angle iron to the radius that you wish for the bend. After you use it, store it away for next time.

Drilling sheet metal is easy using a machinist's drill. Carefully mark the location where you want the hole and make a starting dimple there with a center punch. Remove burrs around the hole using a larger hand-held drill or a file.

Machinist's drills are available in many sizes both numbered and fractional. A no. 80 drill is 0.0135 inch in diameter; large drills over an inch in

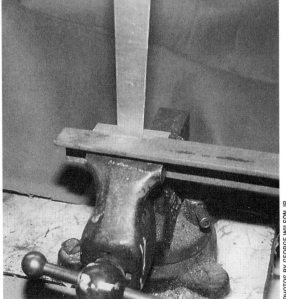
You can make nice bends in a round-jaw bench vise. The jaw shown here was made of a piece of angle iron. Use a wood block and a hammer to avoid hammer marks at the joint. If your vise has rough jaw surfaces, make "soft jaws" from aluminum angle stock.

diameter are also available. These drills work best when used in a drill press but also work well when used in mechanical and electrical hand drills. Large drills should be operated at slow speeds and, therefore, in a drill press. Clamp your work to ensure that the drill does not catch and force the work to rotate or, worse, throw it off the table.

Large, odd-shaped holes can be cut using a series of interconnecting small drill holes and later filing the edges. You can also use a "nibbler"—a tool that takes many small bites (nibbles) of the metal to enlarge and shape the hole. Machine-shop tools, such as shapers and milling machines, can also be used but are usually not available to hobbyists. Other metals generally follow the same rules as aluminum. Brass and steel can be soldered, but aluminum soldering/brazing is tricky and best left to the experts.

Make holes in metal with a hand drill or a drill press. When you use a large drill, clamp the work to the table to prevent it from rotating or being thrown off the table.

A round jaw made of angle iron stock and soft jaws of aluminum angle stock.

This bending tool ensures that the bend radius will be great enough to avoid stress cracks. Use round-jaw pliers or a round-jaw vise to bend thin music wire. You may need to use a hammer to form the bend.

winding tool to form coils like those needed for nose-wheel struts. Thin wire is easily bent using pliers, but again, avoid sharp bends, as they will break when stressed. Round-jawed pliers are helpful.

Thin music wire (and softer wires) can be cut using pliers with cutting blades or with pliers that are specially made for metal cutting. Trim the ends with a fine file or a grinding wheel. Music wire that's $1/16$ inch in diameter or larger is best cut by grinding or filing it about halfway through and then breaking it by hand. Table-mounted or

WORKING WITH WIRE

Most wire used in model airplanes is "music wire," which is really hardened steel. Copper wire is used for electrical wiring and to assist in making soldered music-wire joins. Avoid making sharp bends in music wire. It can be bent around rounded vise jaws or with bending tools. You can use a coil-

Here a 10-32 tap is used to thread the holes for the nylon wing hold-down bolts. The tap is in a T-handle. These handles make it easier to tap holes in less accessible places. Harden the threads in wood with thin CA.

BUILDING BASICS

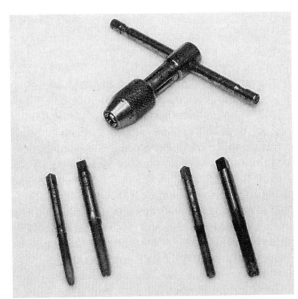

Regular taper (left) and bottoming taps (right). Start the threading with a regular taper tap and finish with a bottoming tap.

hand-held grinders will do the trick. The ends can be very sharp, so be sure to trim them with a file or a grinding wheel. Saws are useless when it comes to cutting most wire.

TAPPING AND THREADING METAL

There are many times when tapped or threaded holes are needed. Again, taps come in many diameters and threads per inch. Some popular sizes used in models are:

Size and threads/in.	Diameter (inches)	Tap drill number
2-56	0.089 (3/32)	50
4-40	0.116 (1/8)	43
6-32	0.144 (5/32)	36
8-32	0.170 (11/64)	29
10-32	0.196 (13/64)	21
1/4-20	0.250 (1/4)	7

There are several types of taps. Most used by modelers are regular tapered and bottoming. Regular tapered taps start easily into their prescribed tap holes. Bottoming taps will cut threads almost to the bottom of the tap hole. If you need threads all the way to the bottom of a blind hole, it is best to use a regularly tapered tap first and finish with a bottoming tap.

Tapping is done by advancing the tap using a tap wrench (I prefer the T-handle type) a turn at a time and each time backing it off a half turn to "clear the threads." Use household or motor oil to make the tap work more easily. Do be careful not to force the tap and break it. Removing a broken tap can be very troublesome, and profanity does not help!

If you're tapping a deep hole, you'll occasionally need to remove the metal chips from the hole. Back the tap out and then dump the chips out. A blast or two of compressed air is helpful.

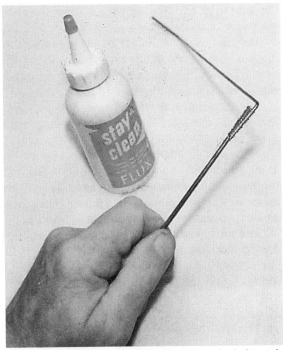

This music-wire joint is wrapped with copper wire before soldering. When the joint is heated (use plenty of heat), the solder will flow into the copper wire and then onto the music wire. Joins made this way very seldom break.

WORKING WITH METAL TUBES

Hard and soft brass and soft aluminum tubes have many uses in models beyond the plumbing for engine fuel. Typically, a metal tube may be used as a bushing inside a wheel axle hole to reduce the hole's diameter. You can easily cut a metal tube by filing (or sawing) it part-way through and then

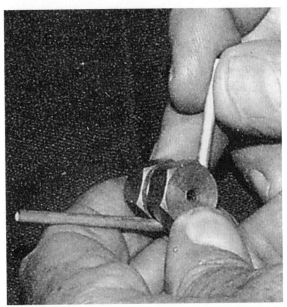

It is very difficult to bend a metal tube without collapsing it unless you use a bending tool. This photo shows a right-angle bend being made using one type of bending tool.

breaking it by hand or with pliers. Trim the ends with a file or sandpaper.

The most difficult thing about working with tubes is making smooth bends. Hobby shops sell tools for doing this; these are well worth their cost and should last for a long time.

Flats and crimps in a metal tube weaken it and invite holes and leaks if it is used in a fuel system.

SOLDERING

Briefly, solder can be used to join steel, brass and copper to themselves or to one another. The metals must be clean (use sandpaper, a file, a scraper, or a wire brush) and, in some cases, you will need a liquid or paste-type flux. You need a strong flux to solder stainless and plain steel.

The soft solders most often used are 60/40 (tin/lead mix) and 95/5 (tin/silver mix). The first is used to make electrical connections and often comes with a core of flux to make it work more easily. This 60/40 solder can also be used for mechanical joints, but the 95/5 solder used by plumbers is stronger and, therefore, a better choice for mechanical joints. The 95/5 solder melts at a higher temperature than 60/40 solder. Silver solder and its flux are available from hobby shops and are even stronger than the other solders mentioned.

All of these solders can be worked with soldering irons. Those that require a torch (frequently called "brazing solders") are harder to use and require special fluxes, but they do produce very strong joints.

The basic trick to soldering (after cleaning and fluxing) is to heat the metals that you are working with. When the metals are hot enough, they will melt the solder and it will flow into the joint to make a strong bridge. Obviously, the iron must be large and hot enough to accomplish the task.

Working with metals is a great way to expand your model building beyond the limitations of buying available parts. You are free to "do it your way," which is the very essence of scratch-building.

2
Landing-gear essentials

WORKSHOP SECRETS 51

Landing-gear ABCs
by George Wilson Jr.

Typical bent, hardened, sheet-aluminum main gear. It will be attached to the fuselage using three 6-32 nylon machine screws that will break during a really hard landing.

This section is aimed at novices and will discuss landing gears. The next section will cover skis and seaplane floats. The complications of retractable and scale gear have been omitted with deference to experts in those areas.

Land gear is of three types: tricycle (with a nose wheel), tail-dragger (with a tailwheel or skid) and the single wheel/skid used most often in gliders. The last type is relatively simplistic; it may be somewhat resilient but it is affected by only one rule: that it be mounted ahead of the model's center of gravity (CG) to ensure that the tail will be down when the glider is at rest. In this respect, single wheels/skids are like tail-draggers. Tail skids make ground maneuvering difficult, but swiveling or steerable tailwheels eliminate most of this problem. The swiveling or steerable nose wheel in tricycle gear provides improved pilot forward visibility and, for model airplanes, has the added advantage of being a propeller saver. Steep landing approaches are less disastrous if the nose gear takes much of the shock.

The ideal wheel position is directly under the CG. This allows the aircraft to pivot (seesaw) easily on the wheels and rotate to a positive angle for easy takeoffs. The actual position of a tail-dragger's CG, however, must be behind the main wheels to ensure that the tail is on the ground when the airplane is at rest. Similarly, the CG must be in front of tricycle-gear main wheels to ensure that the nose wheel is firmly on the ground.

Tricycle gear is used on most trainer models, and it is less tricky during ground handling. And don't write off a swiveling nose wheel, which eliminates a tricky steering linkage and still allows fair steering with the throttle and air rudder. Alignment of the main wheels does not require much precision, but its principles should be understood. Figure 1 defines the mounting angles of the wheels. The wheels often have a bit of positive camber while flying to allow for the gear's bending outward during and after landing. Ideally, the wheels should be near vertical when the aircraft is sitting on the ground. Camber and toe-in/out are similar in effect. Camber is the inward/outward canting relative to longitudinal axis of the aircraft. Toe-in/out is the canting of the wheels

A typical wire single-strut gear, as shown in Figure 2. The parts of the struts that run across the fuselage act as torsion springs and help smooth out rough landings.

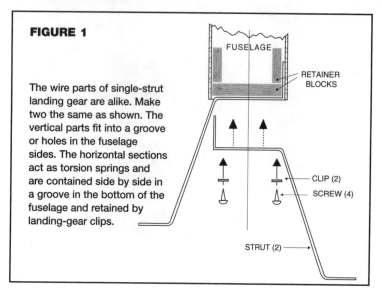

FIGURE 1

The wire parts of single-strut landing gear are alike. Make two the same as shown. The vertical parts fit into a groove or holes in the fuselage sides. The horizontal sections act as torsion springs and are contained side by side in a groove in the bottom of the fuselage and retained by landing-gear clips.

LANDING-GEAR ESSENTIALS

The steering linkage for a steerable nose wheel is shown here. Note the pushrod that reverses itself to provide easy attachment to the steering arm. Freely turning nose wheels also work well with a little practice, using the throttle and air rudder to blast the tailwheel around.

A tailwheel mounted on a bracket at the rear of the fuselage of an Ace R/C Bingo. The tailwheel is steered via a link to the air rudder. A flexible link helps isolate the air rudder from ground steering shocks. Note also the pushrod connection for a water rudder.

Figure 2

Landing-gear angles are defined in this figure. The angles are shown exaggerated to make them more apparent. In the case of models, camber and toe-in/out have similar effects. Toe-in is used to help maintain straight tracking when one wheel lifts. Camber should be adjusted to make the wheels perpendicular to the ground when the model is at rest.

relative to the vertical axis of the aircraft. A degree or two of toe-in is recommended to assist in straight ground running. If one wheel bounces or becomes light when a wing lifts, the drag of the wheel that is touching the ground tends to turn the model toward that wheel's side. With both wheels on the ground, the toe-in of the wheels tends to cancel each other and cause only minimal drag.

Track width is the distance between the main wheels; usually, more is better. The wider the stance, the less the model will tip when wind lifts a wing. When the wheels are close together, the tendency for ground looping increases. This isn't often a problem because it occurs at relatively high speeds when air-rudder control is good. Note that rudder control is used on the ground. Many novices try to use aileron control, which is totally ineffective for ground steering.

Wheel size of functional (vs. scale) models should be chosen to match the roughness of the runway(s) where you will be flying. It's easier to negotiate rough runways with large wheels, but wheels of any size work well on smooth runways. Wheels that have tires with square cross-sections work well on grass runways. The difference between 2- and 3-inch-diameter wheels in shortening takeoff roll is often dramatic. Wheels (and the model's structure) also absorb landing shock. Metal struts are deformed on impact and take on some of the energy, but give it back to the model as they return to their normal shape. Soft tires and true shock absorbers absorb energy, which is then dissipated. If you can't find the type of wheels you need, search the literature for articles on making some.

WORKSHOP SECRETS 53

CHAPTER 2

A typical steerable nose wheel emerging from the underside of the fuselage. The nose wheel should be raked backward to assist it in centering itself.

Tires in many hardnesses are available, from soft inflatable tires to solid hardwood tires. The ones most used on sport models at my field are low-bounce (energy absorbing) and soft-sponge types. Tailwheels should be narrow and have small-diameter tires to ensure good ground control on rough runways. Several steerable tailwheel mounts are available, and some of these include springs for shock isolation. In most cases, the tailwheel is linked to the rudder.

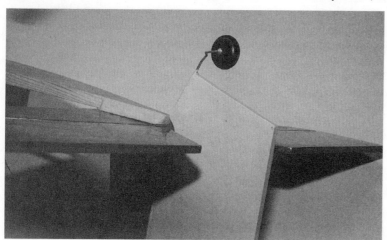

The tailwheel can be mounted directly on the air rudder, as shown here. If you use this mounting method, be sure the rudder hinges are strong enough to withstand the tail weight and steering shocks.

Steerable nose wheels are usually linked to the rudder servo on the side opposite the air rudder pushrod. Again, several nose-wheel mounts are commercially available. It's especially important to provide shock isolation between the nose wheel and the rudder servo because a rough landing can easily strip the teeth off the servo gears. Buy a shock-isolating coupler and mount it on the servo, or bend the nose-wheel pushrod (this also allows another method of adjusting the nose-wheel centering). As mentioned previously, a swiveling nose wheel is simple to make and, with practice, works well. A light centering device is recommended with this method to ensure that the nose wheel is straight ahead during flight. Besides, nothing is quite as pathetic as a model landing with its nose wheel at an angle.

Struts and axles are the most important parts of landing gear. They should be strong and relatively rigid, but flexible enough to ease the landing shocks delivered to the model's fuselage. Sheet-metal landing-gear struts are usually made from one piece of tempered aluminum of thickness suitable for the model's weight. These struts can be attached to the bottom of the fuselage using nylon screws that will shear off during a bad landing. You can make wire struts using two music wire legs on each side and with the cross wires attached to the fuselage bottom with landing-gear clips. I prefer to cut grooves in the fuselage bottom for the cross wires to ensure good alignment and use flat plastic clips that bridge the grooves (the plastic will break during a hard landing). This type of strut can be made by cleaning the wires with sandpaper, wrapping the joints (usually near the wheels) with thin copper wire, applying soldering flux to the joints and soldering using a heavy iron (100 watt is good). To ensure a good fit, mount the strut assembly on the model during the soldering operation.

Single wire struts work very well and are easily made. Two separate struts are used. The fuselage cross wires are side by side in a groove that crosses the fuselage bottom, and they are held in the groove with flat landing-gear clips. The fuselage ends of the struts are bent at a right angle and are positioned in hardwood blocks that are attached to the inside of the fuselage. The cross wires act as torsion-bar springs and help to smooth the landing shocks that are transmitted to the fuselage.

The wheels should turn freely and smoothly. If necessary, use a piece of brass tube as a bushing to ensure a good fit. A wheel collar can be used on the inner side of the wheels so the wheels don't bind on the wire's curve. My favorite solution to this problem is to solder a ¼-inch-o.d. washer to the axle inside the wheels.

Flat metal struts do not have this problem. Several

commercial axle fittings or machine bolts work quite well. Be sure to use locknuts, at least on the outside of the wheels, if you use a bolt to retain the wheels. Wheel collars work best if you file or grind a flat on the axle. Even cup-head setscrews soon work loose if they

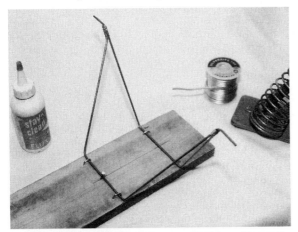

Landing gear that uses more than one strut on a side can be soldered easily by cleaning the wires with sandpaper, wrapping the wire joints with thin copper wire, applying soldering flux/paste and soldering with plenty of heat. Joints made in this way seldom break. Note the jig made to position the struts during soldering.

are against a round surface. A neat way to retain the setscrews is to stretch a short piece of fuel tubing over the collar. This is not a guarantee but works well if the setscrew end is slightly above the collar. Landing-gear-strut drag can be lessened by adding fairings to the struts and pants to the wheels to make them more streamlined. In addition to commercially available pants, many articles and how-to's on wheel pants have been published. Shaped balsa, formed plastic and fiberglass pants all look good and work well. Commercial hardware is available to ensure that wheel pants remain aligned after landings and takeoffs.

LANDING GEAR CONSTRUCTION NOTES

• Cut music wire by notching it with a file or a grinder and breaking it off. A Dremel or a similar hand grinder works well.
• Bend wire and aluminum sheet with generous radiuses to minimize the stress in the material. Use a wire bender (K&S makes a good one) or soft-jaw inserts in your vise.
• Cut sheet aluminum with a bench saw using a carbide-tipped blade. Wear safety glasses.
• Smooth the edges of aluminum with sandpaper. Wire ends can be smoothed with a fine file or a grinding wheel.
• If you lack the confidence to work with wire and/or sheet metal, buy secondary parts from a company that sells parts for its kits (check out Goldberg and Sig).

Landing gear is a very important part of a model airplane, whether it's a single wheel or a scale copy of the real thing.

CHAPTER 2

Ski and float landing gear
by George Wilson Jr.

In the June 1998 issue of *Model Airplane News* (page 92), I starting talking about basic landing-gear layout. This time out, I'd like to finish the discussion by covering the rudiments of skis and floats. These types of landing gear extend your flying to the lake and the winter—including frozen lakes.

Twin floats and skis may be substituted for wheeled landing gear with great results. Contrary to what you may have been told, when properly installed, well-designed floats and skis require little (if any) added engine power. Bear in mind that most land planes are over-powered for aerobatic capability, and that seaplanes and ski planes are not intended to do aerobatics.

Landing gear works most effectively when it makes the airplane act as a balanced seesaw. When flying speed has been reached, the airplane rotates about the landing gear axle to achieve an angle of maximum lift. Landing gear (including skis) have to be unbalanced enough to assure that the nose wheel or tailwheel stays solidly on the ground when the plane is at rest, and in the case of a tail dragger it has to be unbalanced enough to minimize nosing over during taxiing. Seaplanes should be balanced so that the step of the float(s) is directly below the CG, and takeoff rotation occurs at or about the step with minimum effort.

SKI PLANES

Skis can be set up like land gear, with either a steerable nose or a tail ski, and will work quite well. The most serious deterrent to flying with skis—other than the cold weather—is footprints in the snow. A plowed area or, better still, a frozen lake surface make good places to fly from.

Conventional skis set up with nose or tail skis follow the same rules for placement as their wheeled equivalents. They should be large enough to even out the surface irregularities. The larger the skis, the

This is Joel Chappell's PT-40 on Du-Bro "tricycle" aluminum skis. The front ski is used for steering just as a nose wheel would be. The tail-dragger arrangement with skis also works well.

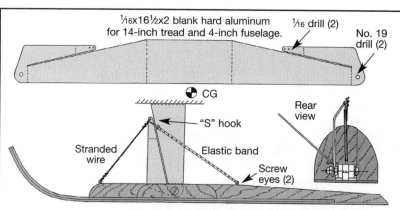

Figure 1. A typical ski-rigging system that works well for gears with a nose ski or a "two point" ski setup that uses neither a nose nor tail ski. If your gear uses a tail ski, the wire and elastic band positions should be reversed.

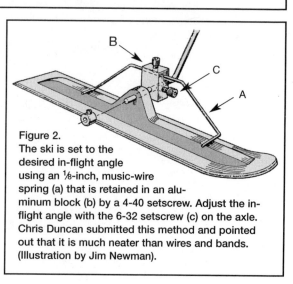

Figure 2.
The ski is set to the desired in-flight angle using an 1/8-inch, music-wire spring (a) that is retained in an aluminum block (b) by a 4-40 setscrew. Adjust the in-flight angle with the 6-32 setscrew (c) on the axle. Chris Duncan submitted this method and pointed out that it is much neater than wires and bands. (Illustration by Jim Newman).

LANDING-GEAR ESSENTIALS

more easily they can bridge unevenness in the snow. For the bottom surface of the skis, a good, hard finish such as polyurethane is recommended. To lessen friction, you can also use alpine ski wax on the bottoms of your model's skis.

Skis should be "rigged" so they remain tilted upward somewhat at their front ends while in flight. This helps prevent them from digging in when the model lands. This is usually done with a spring or rubber band that pulls up the front of the ski. A cord attached to the rear of the ski limits the amount of tilt. During takeoff, the rigging of the skis must allow the model to rotate to allow the wing to achieve the proper angle of attack for takeoff. During flight, the positive ski angle should be small enough to minimize drag but large enough to prevent the tips from catching during landing.

CONSTRUCTION

Skis may be made from ⅛-inch aircraft plywood with a hardwood top center piece to improve rigidity. The top center piece is also used to connect the ski to the landing gear axle. The front of the skis should be generously curved upward to help them navigate over rough snow. This curvature is usually made by soaking the plywood in water and then clamping them into a wooden mold. Metal (usually aluminum) may also be used to make skis. To promote straight taxiing and minimize side motion, a thin keel (usually a narrow strip of hardwood or plywood) can be added to the bottom centerline of the skis.

An example of "two point skis" (no nose or tail ski required). This type of setup depends on the air rudder for ground steering. However, the air rudder can be very effective on slippery surfaces.

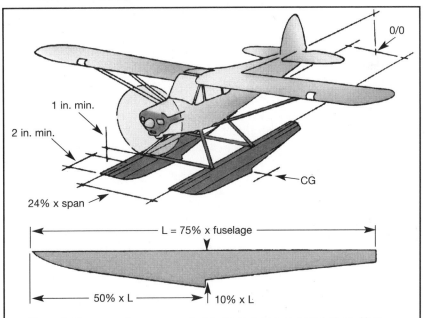
Figure 3. The general arrangement of floats installed on a land plane. Note that the top of the floats are parallel to the fuselage's reference line. Rotation for maximum takeoff lift is governed by the float height at the step and the after-body's bottom angle.

Floats should be rigidly mounted. This one has "N" struts on each side and strong pine cross struts.

On smooth, crusted-over snow and on ice, the ease with which a ski-equipped model can weathervane into the wind takes some getting used to. Steer the model in the direction you want to go, even if its nose is not pointed that way. As usual, taking off and landing into the wind make the weathervaning problem go away.

SEAPLANES

The basic design requirements

WORKSHOP SECRETS 57

for floatplanes and flying boats are the same. They both have the advantage over land planes of having their takeoff rotation points (tip of the step) directly under their CG balance points. As can be seen in the drawings, the float or hull bottom tapers upward both forward and aft from the step. The rear taper must be enough to allow the wing to achieve its proper angle of attack for takeoff when the model is running on the step. The forward taper assists the model to "get on step" during takeoff. In most cases, ground landing gear and floats can easily be interchanged.

Joel Chappell's Headmaster Sport 40 on Northeast Aerodynamics flat-bottom floats. Most trainers make great floatplanes. Note the simple conversion from tricycle land gear.

Flying boats have distinct advantages. First, they deflect the water outward and away from the propeller—assuming they have a single engine. Second, they have ample room to make watertight compartments for your equipment. With most flying boat designs, the wing is close to the water and the airplane is less susceptible to crosswinds, which makes water handling easier. Given a choice, flying boats get my vote every time because of their better landing and takeoff characteristics.

SIZE

Floats must be large enough to adequately support the model at rest. The rule of thumb is:

Float length =
75 percent of the fuselage length.
Float height (at the step) =
10 percent of float length.
Float width (at the step) =
float height at the step.

Ideally, at rest, the nose of the float should be out of the water, and the bottoms of the aft ends should be just about at water level. Flat-bottom floats are easier to build and have increased volume, and that increases their ability to support the model. The rules for flying boats are similar; they must be large enough to handily support the model's weight. Flat bottoms and ample width are helpful features.

CONSTRUCTION

Hulls and floats can be built from balsa/ply, fiberglass, or glass-covered foam. Rather than trying to design your first set of floats (or flying boat), start by building a proven design. There are also many float sets available in kit form.

In any case, floats should have strong forebottoms to withstand beaching or striking hard objects, and their chine edges should be sharp to direct the water sideways. A fiberglass finish is recommended. I have tried many bottom shapes: V-shaped, inverted "V" (viper), curved and flat; they all work well. But, flat bottoms are hard to beat for ease of construction and performance.

WATER RUDDERS

These are most important. Much of the fun of water flying is being able to taxi into the wind for takeoff and then back to the beach when you are done. Occasionally, you will hear someone say his model doesn't need a water rudder. If you watch him fly, you will find he doesn't do much taxiing. Most often the model is released

Harry Newman's de Havilland Beaver on landing approach; a great floatplane. Note the deployed flaps being used during approach.

Scale floatplanes are very popular. Here is Fred Tuxworth's OL-6 Loening Duck that has functioning amphibian gear. It taxies into the water, retracts the wheels and flies away. To land, it reverses the procedure and taxies back up onto the beach.

at full throttle and is airborne very quickly.

Water rudders are linked to the air rudder and when the model rises on the step, the water rudders sometimes come out of the water; this is when the air rudder becomes effective. Rudders mounted on a piece of music wire extending from the air rudder hinge line appear flimsy but they are very effective and easy to construct.

The spread between the floats should be about 25 percent of the wingspan to minimize tipping tendency when the wind lifts a wingtip or during tight turns while taxiing. Most flying boats have wingtip floats to minimize tipping.

Mounting struts for floats should be rigid. Struts should be X-braced lengthwise and similarly cross-braced, if possible. Struts between the floats are also recommended to increase rigidity. Floppy floats resist getting up on the step.

Floats should be mounted so they cause minimum drag in the air. Their top surfaces should be slightly

Cubs love to be seaplanes. Here is Sonny Martel's ¼-scale Cub settling in at the Southern NH Flying Eagles seaplane meet.

negative. However, if they are poorly designed and you can not get enough rotation for easy takeoff, they may be shimmed to give the wing a more positive angle; this usually solves takeoff problems.

CHAPTER 2

Install landing gear in a foam wing
by Les Morrow

There are many ways to install landing gear in foam wings, and most work pretty well. If you aren't careful, though, overzealous bracing can add a lot of weight. It also seems that if something is easy to install, it just doesn't last very long.

Over the years, my son and I have built and flown warbird racers and have sometimes had trouble keeping gear mounts in the wing, especially with the finished weights of the aircraft we race. We tried all the methods we had read or heard about and have

This Yak 3M designed and built by the author uses Robart retracts and the type of landing-gear mounts described here.

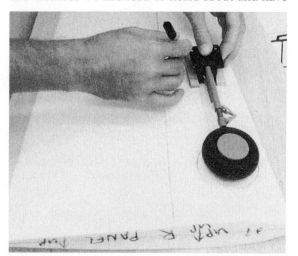

1 Lay out the retract unit on the wing-core before sheeting and mark the position of the mounting plate.

had limited success installing conventional landing gear in WW II single-engine fighters, but occasionally, a landing-gear mount would loosen or fail on a less-than-perfect landing. We tried various ways to securely mount the gear and still keep the weight reasonable, and we figured out a foolproof installation that's secure, light and compact, easy to install and will take all the abuse we hand it (short of an all-out crash). If you're interested in a landing-gear mount that will take lots of abuse, will still function correctly landing after landing and doesn't

weigh as much as the engine, keep reading.

The first thing to do is make ¼-inch-thick birch aircraft plywood landing-gear plates to fit your retracts. The plates need to be bigger than the gear assembly by about ¾ to 1 inch all around. After you make the plates, temporarily mount the

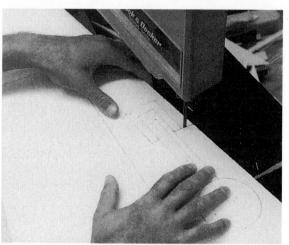

2 Cut ⅛-inch slots for the gear-mount ribs, then cut out the wheel well.

gear to each plate so you can fit the plate into the foam wing-core (before sheeting) and align the rake, castor, retraction angle and position of the gear on the wing.

Mark the wing-core and carefully remove foam as needed to clear the wheel, strut, gear retract mechanism and mounting plate.

After the proper amount of foam has been removed and you are satisfied with the position, rake, castor, etc., take the core to your saw and cut two 1/8-inch-wide slots from the leading edge of the wing, chordwise, to approximately 1/2 inch past the hole you made for the mount. Make a cut on each end of the

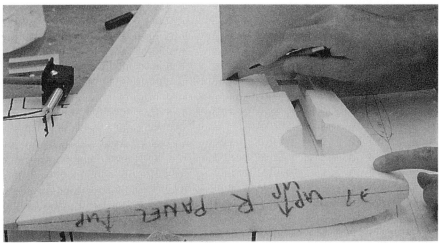

3 Insert 1/8-inch ply and mark the airfoil to fit the wing-core.

mount. I know; it looks as if the wing is going to fall apart, but trust me; it will all work out. Now take a sheet of 1/8-inch birch aircraft ply (3-ply minimum; 5-ply is better) and slide it as far as you can into the slot you just cut. Next, draw around the core on the plywood (you're marking a landing-gear rib that will fit the wing profile exactly). After you've drawn around one plywood rib, slide the gear mount (minus the retract unit) in place tightly against the 1/8-inch ply, then draw a line on the plywood rib to mark where to cut a slot to insert the mount into the rib. Remove the new rib and gear mount, then repeat for the other end of the mount. When you've finished, do the same thing for the other wing.

Cut out your plywood ribs, check the fit of all parts and epoxy the entire landing-gear mount assembly into the wing with 5-minute epoxy. Tape it into place and set it aside until it has dried.

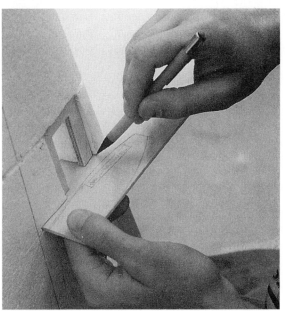

4 Temporarily insert the mounting plate to the position marked on the rib blank.

When the glue has dried, check the position of the plywood ribs and sand them flush with the foam wing surface, if needed. You probably won't need much sanding, if any, assuming you did a good job of test-fitting the assembly on the bench before inserting it into the wing.

After you place the other controls, e.g., the pushrods and aileron/flap bellcranks, in the wing, prepare the wing for sheeting as per your favorite method. Sheet the bottom surface of the wing-core first so you can see where to make the wheel well and strut cutouts. If you're a purist,

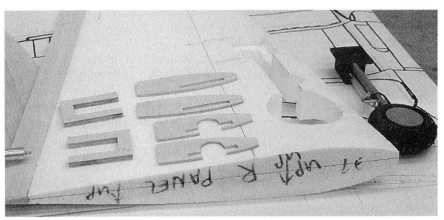

5 The landing-gear mount parts have been cut out and are ready for installation.

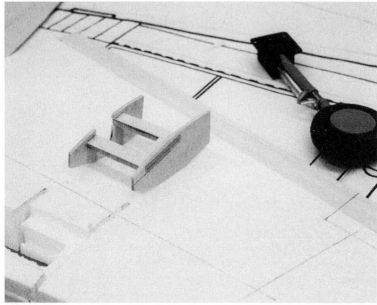

6 The subassembly is ready to be inserted into the wing-core. Epoxy it in place and hold it with tape until it dries.

7 When everything has dried, sheet and sand the wing-cores.

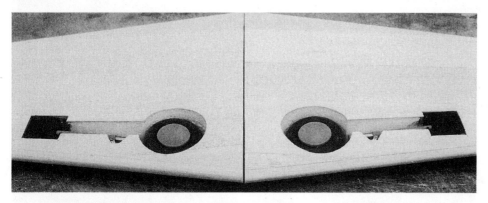

8 The finished foam wing with the landing-gear mounts in place. The wing needs only to be covered.

prepare a wheel-well liner of thin balsa, 1/64-inch ply, or something similar. Because we race our warbirds around pylons with other aircraft on the same course, and the potential for a midair is always present, we just sand the foam wheel well carefully, coat it with epoxy, then prime and paint. This method works well and is quick and light. Once you've finished the cutouts for the wheels, struts and retracts, finish sheeting the wing as usual, attach your leading and trailing edges, and sand to finish. After the wing is finished and the gear has been installed, build and paint gear doors and retract covers to match your documentation and color scheme and attach them using your favorite method.

This may sound somewhat complicated, but study the photos carefully. You will find this method is quicker than reading this article, and the mount assembly will be strong enough to outlast the airplane. You'll never go back to your old way of mounting retracts once you find how light, easy and strong this method is; we sure haven't!

LANDING-GEAR ESSENTIALS

Install shock-absorbing Oleo landing gear
by Gerry Yarrish

The next best thing to having a model with retracts is having one with functional Oleo-strut landing gear. The shock-absorbing abilities of these struts improve landings immensely, and the improvement in scale appearance is unquestionable.

Robart Mfg. has been making Robo Struts, retracts and custom struts for more than 15 years. When I was building my latest model—the Ohio RC Chipmunk, I wanted scale, functional landing gear. I wanted them to slip over the wire gear already installed in my wing, so I gave Robart a call. Here's how to install custom struts on your next project:

1 Before you begin, you must determine the distance between the bottom of the wing and the axle and also the diameter of the landing-gear wire to which the struts are attached. The landing-gear wire on my Ohio RC Chipmunk is ¼ inch.

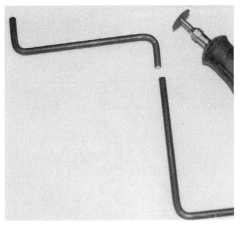

2 Clamp the landing-gear wire in a bench vise. Measure down the landing gear wire, and mark off a length of about 1 inch. Cut the wire with a cut-off wheel chucked in a Dremel* Moto-Tool. Don't forget to wear safety glasses!

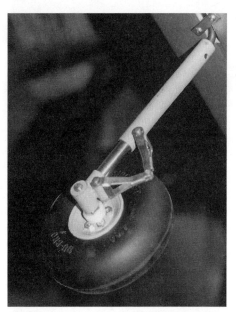

3 Slip the struts over the landing-gear wire, and snug the setscrews that hold them on the wire gear. Don't tighten them yet.

WORKSHOP SECRETS 63

CHAPTER 2

4 Remove the wheels and the axles from the struts, and place the aircraft so that the struts are on a flat surface. Use a long, straight, aluminum L-angle to align the lower struts' axle fittings with each other. Now twist each strut inward slightly to achieve approximately 1 to 1.5 degrees of toe-in for each axle. This greatly improves ground handling by reducing your model's tendency to ground loop. Now tighten the four setscrews at the top of the struts.

5 Remove the struts, and grind four flat spots about ³⁄₁₆-inch wide where the setscrews have made indentations on the wire.

6 Reinstall the struts, check their alignment and toe-in adjustments again, and tighten the setscrews but, this time, add blue Loctite. Tighten the setscrews firmly, and check the axle alignment once more.

Put the wheels onto the axle bolts, slip a Du-Bro nylon washer on, and thread on the thin-profile locknut. Add Zap Z-42 Thread Lock to the axle bolt, and screw it into the strut's axle fitting. When you have about ¹⁄₃₂ inch of clearance between the bolt head and the wheel, tighten the locknut while holding the axle bolt with an Allen wrench so that it doesn't rotate. When the Thread Lock is dry and the landing-gear wire is secured to the wing, you're ready for some improved ground handling. Custom struts look great and are easy to install, give Robart a call for your custom landing gear needs.

LANDING-GEAR ESSENTIALS

Make custom landing-gear clips

by Randy Randolph

Torsion-bar landing gear is standard equipment on almost all wing-mounted setups. To mount this type of gear externally, just drill a hole in the plywood plate and install landing-gear clips that match the size of the wire that's used for the gear. Clips are available to fit common wire sizes, but often, it's hard to find the right size. Here's how to make landing-gear clips for any installation.

1 Using sheet-metal shears or heavy-duty scissors, cut a 1-inch-wide strip from the side of a tin can. Mark off ¼-inch strips with a scribe.

2 On every other line, measure ⅛ inch from each end of the strip. Mark the holes with a center punch or scribe, then drill 3/32-inch holes.

3 Cut the clips along the lines as shown. You now have clips that are 1 inch long and ½ inch wide with holes that are ¾ inch apart.

PHOTOS BY RANDY RANDOLPH

4 Open your vise jaws to about 1/32 inch wider than the width of the landing-gear wire. Center the clip over the opening between the jaws, then lay the wire on the clip directly over the opening. Use a hammer to shape the clip.

5 Flatten the wire and the clip so that they're flush with the vise jaws. Two hardwood blocks can be substituted for the vise and will work equally well (possibly even better). A minimum of two clips is necessary for each landing-gear leg.

6 Round off the square edges of the clips. If the wire is larger than ⅛ inch in diameter, the clips should be at least ¾ inch wide.

CHAPTER 2

Make concealed axles

by C.H. Bennett

After I had completed a good-flying Ford Tri-Motor model, I took a long look at the main gear wheels, which had functional—but somewhat ugly—exposed wheel collars. I wanted the finished wheels to have slightly domed, "baby moon" wheel discs like those on full-scale Tri-Motors. This article explains what I did to improve the appearance of the wheels.

Essentially, this method involves fastening the wheel assembly using one wheel collar mounted to the axle, inboard from the wheel. The setscrew in the wheel collar goes through a brass tube bushing and is locked into a flat that has been ground into the axle. A suitable low-profile, rounded screw head is on the outside of the wheel, and it can't come loose, as it is soldered into the brass tube. You can then add a hubcap to cover the outside of the wheel.

1 In this instance, I used a 5/32-inch axle with a typical 2¾-inch-diameter foam tire. (You can also use this method with 1/8- and 3/16-inch music-wire gear axles.) Here, I use 3/16-inch o.d., 5/32-inch i.d. brass tube and no. 8 thread, low-profile brass screw heads (terminal screws from household receptacles and switches). I cut off the surplus thread to leave about 3/32 to 1/8 inch under the screw head. I sweat-soldered the screw heads in place using a suitable length of the brass tube. (It helps to use a longer length of tube while you're soldering it.) I solder both ends on before making the final cut. Try to make a good joint without leaving a large fillet of solder, but if a little is left, you can chamfer the wheel bore to compensate.

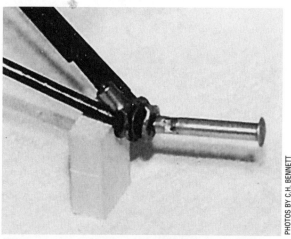

2 Cut the tube to the proper length using a grinding wheel or tube cutter and deburr its i.d. so you can easily slip the tube over the axle. (Determine the length using the sketch, taking into account the wheel thickness and space for thrust washers or side clearance.) Use a felt-tipped pen to mark where you'll grind a flat through the brass tube and into the axle in one motion. Be sure the flat faces in a direction that you'll be able to access later. The flat should be about 1/32-inch deep, and the width must be greater than that of the wheel-collar setscrew. In this case, I made the slot width about 5/32 inch (the 5-40 setscrew has a 0.120-inch diameter). When you pull the tube off the axle, the slot will deburr itself to some extent, but you should touch it up with a small file or scrape it with a no. 11 blade.

PHOTOS BY C.H. BENNETT

LANDING-GEAR ESSENTIALS

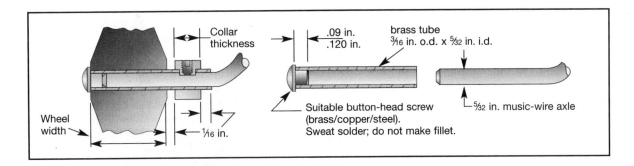

3 Here are the inside and outside of two wheel assemblies. In this case, the plastic wheel bores must be opened to freely accept the 3/16-inch o.d. brass tube. Redrilling these plastic wheels properly is a must to avoid wobbly runout. I do this in a drill press, opening up the bore successively 1/64 inch at a time until the tube fits properly. I drive from both sides of the wheel so that the drill centers well.

4 To dress up the wheels a little more, I epoxied on soda can "baby moon" wheel covers. A word of caution: during final installation, the setscrew must be accurately aligned to the slot in the tube and the axle flat. Wiggle the setscrew judiciously so you can "feel" whether it has entered the tube properly before you tighten it; otherwise, you could easily crush the brass tube. A little light grease or oil will also help everything slide together easily. The idea has worked well for me, and I have made this change to four planes so far.

Make a steerable tailwheel
by Ron Bozzonetti

Using basic tools, it's easy to make a model tailwheel assembly out of music wire and ¼-inch-wide brass strips. This tailwheel is also inexpensive and can be custom-made to fit any aircraft.

Here, I've described the process for making a tailwheel for a .60-size model, but you can make one for a .40-size model if you substitute 1/16-inch-diameter wire for the 3/32-inch shown.

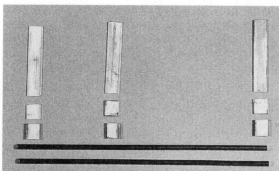

1 Two pieces of 3/32-inch diameter music wire are cut to suit your model (in this example, they are each 4 inches long). The pieces of ¼-inch-square (0.062 and 0.032 inch thick) brass, cut off ¼-inch-wide brass strips, are needed to achieve the same thickness as the 3/32-inch-diameter music wire. The longer piece of brass strip (0.016-inch thick) will be used to form a collar around the assembly.

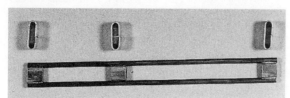

2 The ¼-inch-square brass pieces have been silver-soldered between the 3/32-inch-diameter music wires. The thin brass strips have been formed into a collar to fit over this unit.

3 The formed brass collars have been silver-soldered over the ¼-inch brass pieces.

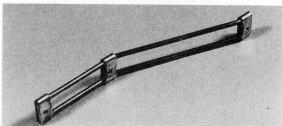

4 The unit shown in photo 3 has been placed in a vise and bent to the desired angles. Holes were drilled (prior to bending) to accept two hold-down screws and the tailwheel bracket.

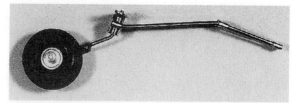

5 The assembled tailwheel bracket ready to be mounted on an airplane.

6 The assembled unit attached to an airplane. A tiller arm was made from brass strip and silver-soldered onto a wheel hub. Springs, omitted from this photo, would be attached to the tiller arm and the brackets would then be attached to the rudder.

LANDING-GEAR ESSENTIALS

Another approach to building a steerable tailwheel
by William R. Nielsen Jr.

I have never been completely satisfied with the standard, steerable tailwheel assemblies available for small to midrange models. They were either difficult to mount or awkward and unsightly. Several years ago, I resolved to find a better way. I believe I have succeeded, and it is a simple arrangement I have used many times.

The system described here is fairly easy to construct and can be adapted to most models during construction. My tailwheel design can also be used with existing models, but you may need to modify the model.

MAKING A DRILLING JIG
It isn't essential, but a drill press is highly desirable for all drilling and, if you plan to make more than one or two tailwheels, you should make a drilling jig. To do this, fasten a 4- or 5-inch strip of ³⁄₃₂-inch-

YOU WILL NEED
- Music wire (use ¹⁄₁₆-inch wire for mid-range models, ³⁄₃₂-inch for larger and .047-inch for smaller models).
- Three small washers with an inside diameter (i.d.) that matches the music wire and a single washer with an outside diameter (o.d.) of at least ¼ inch.
- Brass tube with a matching i.d. and one ¼x¾ pan-head nylon machine screw (bolt).
- A Du-Bro tailwheel with diameter to suit the model (other brands can be used, but I prefer these because of their wide, machined-aluminum hubs).
- Solder and flux (I use Stay-Clean or Stay-Bright).
- A 3- to 4-inch-long piece of ¾x¼-inch plywood or very hard balsa;
- A short length of Nyrod or a pushrod and clevis.

thick, ¾-inch-wide iron or brass plate to one face of a scrap hardwood block with a screw at each end, in opposite corners. Then, drill the plate for the wire sizes you will use. Use a no. 54 drill for .047-inch wire, a no. 50 for ¹⁄₁₆-inch and a no. 37 or 38 for ³⁄₃₂-inch wire. Make these holes about ¾ inch apart, through the metal and wood. Temporarily remove the metal plate, turn the block over and, with a ½-inch-diameter bit, drill (countersink) the holes about ½ inch deep, then drill ³⁄₁₆- or ⁵⁄₆₄-inch-diameter holes the rest of the way through. Now tap the holes with a ¼-20 tap. Replace the plate, and you have a clever, simple, drill jig that makes it easy to drill nylon bolts. Many thanks to Charlie Brandon for his suggestions for making this jig.

Screw the nylon bolt into the jig, and tighten it so that the drill will not loosen it. Don't rush the drilling, and don't turn the bit too quickly. After you've drilled the bolt, remove it from the jig and make sure that the wire is snug but still can turn. You may need to increase the hole one bit size, but you don't want a sloppy fit!

MAKING THE TAILWHEEL
Cut about a 5-inch (or longer) length of music wire and bend it to the desired yoke shape. An inch or so above the top of the yoke, bend it back 15 degrees from vertical. This will provide good tracking. This

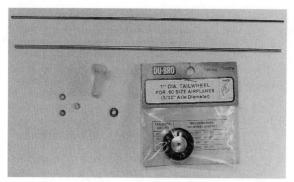

The materials required (other than solder) are a suitable tailwheel, music wire, brass or aluminum tube, four washers and a nylon ¼-20 bolt.

The bolt-drilling jig. Note the holes for three sizes of music wire; underneath, the holes are countersunk to accept the nylon bolt.

The tailwheel components are ready to be assembled.

uppermost bend is where the large washer and one smaller washer will be soldered to create a bearing surface against the head of the nylon bolt. Now bend one end of the brass tube 90 degrees, forming an arm that's more or less ½ inch long. Do not bend it at a hard angle, but rather with a radius, as shown in the diagram. Flatten the end of the arm horizontally and drill a hole to suit the type of clevis you use. Now cut the tube off about ½ inch below the bend. Put the wire through the nylon bolt, and cut the bolt so that when you place it on the wire, the tube (steering arm) will just clear the bolt (about 1/32 to 1/16 inch of clearance will be fine).

Remove the wire and solder the upper end of it (the bend above the yoke and both ends of the

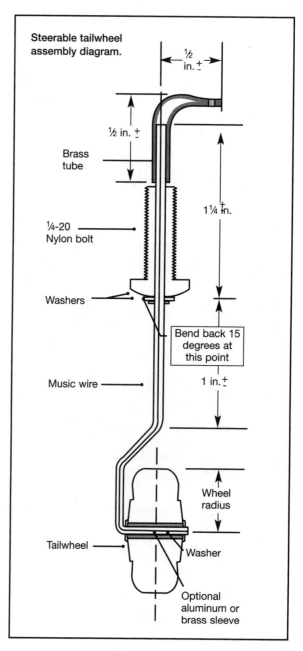

Steerable tailwheel assembly diagram.

LANDING-GEAR ESSENTIALS

Installed and ready to go!

axle); don't use too much solder. Also put a small drop of flux in the end of the tube that slips over the wire and tin the tip. Replace the wire through the bolt. Put the steering arm on the wire at 90 degrees to the thrust line. Now, quickly solder with a minimum of heat to avoid damaging the nylon bolt. Solder a washer at the inboard end of the axle. If you're using wire of a smaller diameter than the hole through the tailwheel, you may need to use a brass or aluminum sleeve on the axle. This should be about 1/32 inch longer than the thickness of the wheel hub. Slide the tailwheel on and solder the outer washer to the axle. Go easy on the solder! Drill and tap the 3- to 4-inch piece of plywood or hard balsa for the 1/4-20 bolt and mount it into place in the bottom of the fuselage, making sure it is far enough forward so that the steering arm clears the sides and any interior structures. Ease the steering arm through the bolt hole and screw it tightly into place. If your model's fuselage is made out of sheet balsa or plywood, you may have to cut an access hole for the steering arm. Use a 1¼- or 1½-inch washer as a guide for your knife and, after connecting the clevis to the arm, patch the hole with covering material. This is easily removed if the tailwheel accidentally becomes disconnected or needs to be replaced.

Now connect the Nyrod/pushrod to the steering arm and the rudder servo, and the project is complete! Use about half the throw you have on rudder, and you will have very responsive yet controllable steering.

Make giant-scale skis
by Roy Vaillancourt

Roy Vaillancourt shows off the ¼-scale, ski-equipped Stinson L-5 Sentinel, which he designed. It's powered by a Quadra Q-42.

While reading a past issue of *Model Airplane News*, I came across an article about float flying off water. It started me thinking about how much fun it would be to fly off snow with skis. After working out some details about writing this article with associate editor Gerry Yarrish, I started.

First, I had to pick some suitable subjects to modify for ski installation. That was the easy part, because my Stinson L-5 Sentinel and Cessna L-19 Bird Dog were just begging to get out of winter storage and be drafted back into service. They're both ¼-scale tail-draggers and are very suitable for trudging through snow.

SKI DESIGN

During a quick look through some full-size-aviation magazines, I came across a short article about winter flying—with skis. This article contained some neat color photos of two Piper J-3 Cubs—probably the most common aircraft—outfitted with different brands of ski, and this supplied me with a few ideas on designing a simple yet effective set of skis for my ¼-scale models.

I generated full-size drawings for the skis following the tried-and-true "That-looks-about-right" formula (good old eyeball engineering!). After measuring the skis and fuselages in the photos, I calculated their comparative lengths, and I used these figures to plan the dimensions of my skis. The length of the skis should be approximately 50 percent of the fuselage length, and the axle pivot point should be 30 to 40 percent of the ski length aft of the ski nose.

MATERIALS

The materials used for the skis are well-known by all modelers and, depending on the weight of your model, the skis can be made of ⅛-, ³⁄₁₆- or ¼-inch-thick lite-ply or lauan (the plywood material used to skin interior household doors). For models that weigh up to about 15 pounds, use ⅛-inch-thick material; models of 25 pounds or more can take ¼-inch-thick material.

Metal skis mean trouble because snow really likes to stick to cold metal. Wooden skis work better; just be sure that you sand their bottoms silky smooth, seal them well with polyester resin, polyurethane, etc., and apply wax. (We've successfully used beeswax as well as high-grade automotive paste wax.)

The center stiffener and the two axle mounts are made of various types of plywood. For ¼-scale models, the stiffener is ½-inch-thick exterior-grade house-construction plywood, and the two axle mounts are ¼-inch-thick aircraft plywood.

Here are all the wooden parts for one set of skis (see text for details).

LANDING-GEAR ESSENTIALS

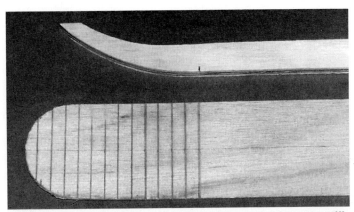

To bend the front of the ski upward to match the curve of the center stiffener, a series of cuts is made in the ski's top surface.

Here, all the parts are clamped together while the adhesives dry.

A finished set of skis, ready for painting.

The entire assembly is glued together with 12-minute epoxy and then painted with a couple of coats of paint and topped off with some clear polyurethane or epoxy.

FUSELAGE CONSIDERATIONS

One of the neatest things about this design is the ease with which you can switch from wheels to skis. This is very important when you get that unexpected snowfall and last-minute calls from your flying buddies to meet them at the field. It takes only a few minutes to change from wheels to skis.

Only one modification is needed: install two pairs of eyehooks on the fuselage to act as attachment points. Install two in front of the landing gear—one on each side. You'll attach the skis' nose bungees to these (more on this later). The other two go aft of the landing gear; the rear extension limiting cables will be attached to them. I simply epoxied some hardwood blocks inside the fuselage and permanently screwed the eyehooks into place (see photos).

SKI SETUP

To set up your skis properly, there are two basic yet important alignments to maintain.

• Toe-in. The skis must be parallel to each other as well as to the fuselage centerline (a function of the landing gear's axle toe-in adjustment).

• Angle of attack. The skis' angle of attack must be approximately 10 degrees positive while the aircraft is in flight (a function of the bungee and limiting-cable adjustments. The nose bungees are big rubber bands that lift the tips of the skis. To limit how high the ski noses rise, you adjust the lengths of the rear limiting cables. I like to make these adjustments on the workbench with the skis mounted on the axles (held in place with wheel collars) and the airplane's tail propped up. To get the required 10 degrees of ski nose-up in a flight attitude, I keep the skis flat on the bench and raise the tail so that the plane's nose is set at a flight attitude of negative 10 degrees.

When you lift the model off the bench, if you've set everything up properly, the bungees will lift the noses of the skis and make the limiting cable taut. When the model is placed on the ground, the cables should slacken and the skis should lie flat. It's important that they also be able to pivot freely on the axles.

To make it easy to attach the bungees and cables, I install line connectors or some other type of "quick-disconnect" device at the fuselage attachment points. Old control-line connectors work well; you might find similar connectors in a fishing-tackle store.

To make it easier to remove the wheels from my models, I've replaced the usual wheel collars with cotter pins that go into small holes drilled through the end of the axles.

TIPS ON SNOW FLYING

With all the shop work finished, it's time to head for the field. The toughest part is waiting for the snow! I live on Long Island, NY, and we don't usually get much snow, but last winter, the snow made it difficult to get to the field! When the Snow Bird Squadron—John Julian, Mike Gross and I—at last

WORKSHOP SECRETS **73**

CHAPTER 2

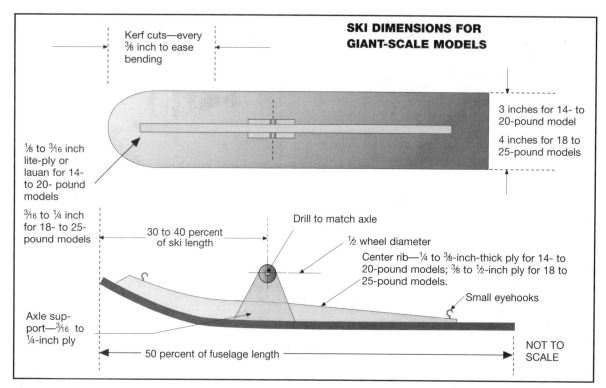

managed it, we enjoyed really great, off-ski flights.

When flying off snow, remember these tips:

• Apply slightly more power to taxi. If you have no ski attached to the tail wheel, the rudder will also need a blast of power for turning.

• You'll need more power for takeoff, and the skis will have to plane on the snow before you'll be able to build up air speed. To overcome torque, apply the throttle gradually and smoothly (just as if you were flying off a green runway). Equipped with skis, your model will not fly as fast because skis increase drag.

• Increase power during landings, and use a slightly nose-high, 3-point, or wheel-landing, approach to keep the tips of the skis up. For short-field operations with my L-5, I particularly like the "I have arrived, 3-point, plop-type of landing."

Using scale snow skis is a really easy way to extend your flying season. Just dress warmly and take along some hot coffee; you don't want frost-bitten flying thumbs. Enjoy!

After the paint has dried, filler pieces are added to and between the uprights so that the attachment point is the same width as the axle's original wheel hub. The cotter pin can be removed quickly—very convenient.

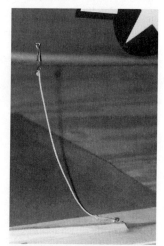

The rear limiting cable is really 50-pound test mason line; it's attached with metal eyehooks and a control-line connector.

The nose bungee is a thick rubber band that's attached to the ski's nose and also to the fuselage with a metal eyehook. The line connector is from my control-line flying days.

LANDING-GEAR ESSENTIALS

Make soda bottle snow skis
by Elson Shields

The nose bungee is a thick rubber band that's attached to the ski's nose and also to the fuselage with a metal eyehook. The line connector is from my control-line flying days.

Are you missing winter snow-flys because aluminum skis are too expensive? Do wooden skis require too much time to make? Are you tired of waiting to fly until the snow is packed down? Well, a 2-liter soda bottle, a trip to the scrap-wood box, a little glue, two steering arms and a couple of evenings are all that stand between you and great winter flying with a .40-size airplane.

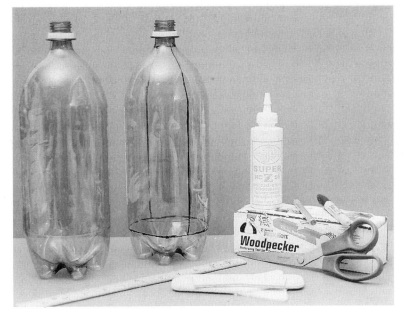

1 You'll need very few things for this project: a 2-liter soda bottle (two, if you are prone to cutting errors); flexible, waterproof glue such as JZ Products' RC-56 (epoxy will not work); a few pieces of lite-ply from the scrap box; and a couple of steering arms. The necessary tools include a ruler, a felt-tip pen and something to perforate the plastic bottle (I used a Top Flite Woodpecker).

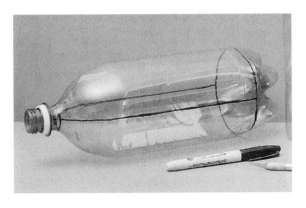

2 Examine the soda bottle and find the two mold seams that run vertically along its length. Mark them with a felt-tip pen. In addition, mark a line around the base of the bottle where it gets thicker. The bottle in the picture has the cutting lines marked on it. Begin by cutting off the bottleneck where it widens. The material here is very thick; you will need a hobby saw or hacksaw. Next, use a sharp hobby knife or scissors to remove the bottom of the bottle. It's easier to cut this material with sharp scissors than it is with a knife. Then, simply cut along the vertical seams, and you have two skis.

WORKSHOP SECRETS **75**

CHAPTER 2

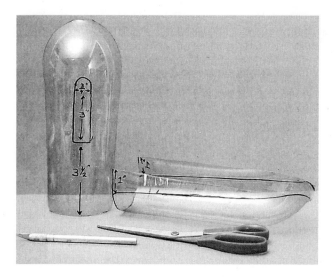

3 The bottle on the right is an example of the rough-cut ski. I angled the rear of it downward to improve its looks. Note the line on the ski that shows where to make your cuts. Mark the area where the lite-ply plate will be glued, as shown on the left ski. The size and shape of this lite-ply piece are not critical, but it needs to be centered over the approximate CG of the ski.

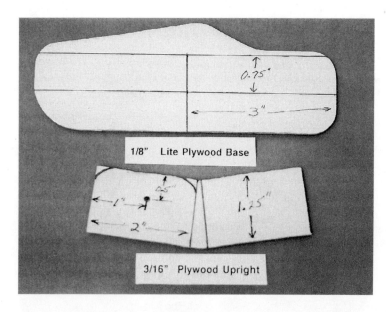

4 The ski is attached to the landing gear with a plywood pylon. Cut the two, ¾x3-inch bottom plates out of ⅛-inch lite-ply. Remember to cut the bottom piece with the grain running lengthwise for maximum strength. The two vertical brackets are cut out of ³⁄₁₆-inch plywood because lite-ply does not have the strength needed for the vertical bracket. Cut these 2x1¼-inch pieces with the grain running lengthwise.

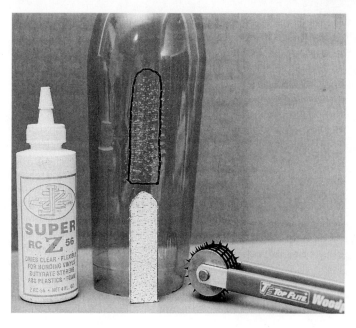

5 Before gluing the lite-ply base to the ski, perforate the area of the ski to which you will be gluing the base. Using the Top Flite Woodpecker, I placed the ski upside-down on a small box and perforated the gluing area from the outside of the ski to increase the gluing surface. To allow the glue to form a stronger bond, the ply plate needs to be roughened; I used the Woodpecker here as well. Use a flexible, waterproof glue to attach the ply plate to the ski. RC-56 worked best, but similar glues will also work. After applying a liberal coat of glue, clamp the plate to the ski and allow it to cure overnight. The most effective way to do this is to sit a 6-ounce can of tomato paste on its side on top of the plate, then clamp the whole arrangement to the table.

LANDING-GEAR ESSENTIALS

6 Glue the vertical attachment to the bottom plate with 30-minute epoxy. You can reinforce the glue joint by adding pieces of ³⁄₁₆-inch-square balsa stock to either side. Triangle stock would also work in this application. After the glue has cured, mark and drill the hole for the landing gear. Drill the hole just slightly in front of the ski CG so the tip of the ski will tend to lift during flight. Be careful not to make too large a hole; it needs to be the exact size of the landing gear.

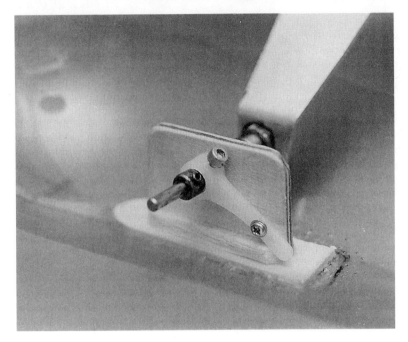

7 Use a steering arm to hold the ski in the proper position. On tail-dragger planes, set the skis parallel to the wing's incidence; tighten the steering arms to hold the skis in place. Add a wheel collar to help hold the ski on the landing gear.

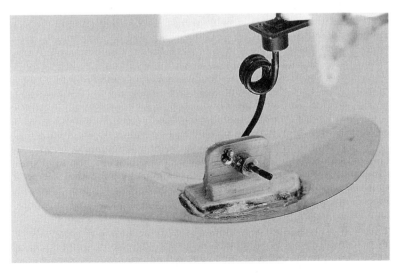

8 You'll need to construct a smaller version of this ski for the tailwheel. Cut it out of the shoulder area of the bottle using the natural bend in the bottle for the nose of the ski. Planes with tricycle landing gear will need a third, full-size ski for the nose wheel. These skis are suitable for most .40-size airplanes. The skis shown have been used for two winters without a problem. The shape of the skis makes them more tolerant of soft, fluffy snow than aluminum or wooden skis. Skis fashioned out of 3-liter soda bottles would be suitable for .60 size airplanes.

WORKSHOP SECRETS 77

CHAPTER 2

Make carbon-fiber landing gear

by Thayer Syme

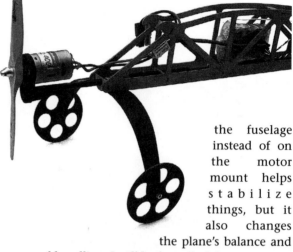

From time to time, a model comes along that does not lend itself to landing gear made with traditional materials. Especially with the very popular slow and park flyers, weight is critical. This prevents you from using heavy anchor blocks for torque wire, multiple pieces of braced wire, or sheet aluminum. When confronted with a situation like this, custom landing gear made out of carbon fiber and epoxy makes a lot of sense. The use of composites is uncommon for landing gear. This is a shame, as it is quite possible to make stronger and lighter landing gear with these materials. Let's look at a typical park flyer to trace the steps of making and installing carbon-fiber landing gear.

The Diversity Aircraft Dragonfly is a 48-inch-span, 450-square-inch foam-wing park flyer that has a direct drive S-400 for power. Its flying weight is 15 ounces. The original landing gear was made out of music wire sandwiched between two layers of plywood in the motor mount. The gear would often split the mount because of the inevitable bumps and bruises of landings while pilots learned how to fly. The gear's excessive flexibility also allowed the wings to drag while landing or taxiing.

To avoid these problems, I decided to try a new set of vacuum-bagged, carbon-fiber gear and mount it on the carbon-fiber fuselage backbone and plywood frame. Mounting the new landing gear on the fuselage instead of on the motor mount helps stabilize things, but it also changes the plane's balance and ground handling. I will look at all of these concerns as I complete the project. Vacuum-bagging eliminates the need for any excess resin, making the part even lighter and stronger.

THE MOLD

Make a very simple mold using a piece of scrap pine 2x4. Keeping the intended track and height of the landing gear in mind, I made my mold 3 inches tall and 8 inches wide. I cut the edges of the block by making a gentle arc from the midpoint on the top to the bottom on each side. After sanding the block, I covered it with MonoKote, which does an excellent job of sealing the mold's surface and also acts as a release agent for the carbon fiber. Even

 Safety first!

Whenever working with composite materials, common sense should rule the day. For personal safety while doing this or any project involving epoxies with carbon fiber or fiberglass, I use and recommend that you use a barrier cream on your hands as well as a pair of thin rubber gloves. A respirator and a fan are also a good idea to help you avoid inhaling epoxy fumes or stray fibers.

Unlike wood splinters, bits of carbon fiber or fiberglass are difficult, if not impossible, for your body to recognize. Triggering of natural defense mechanisms is therefore much less likely to occur. Cleanup is best handled with white vinegar followed by soap and water. I have read enough horror stories about overexposure to epoxies as well as inhalation of sanding dust to ensure that I go out of my way to take these simple precautions. I suggest we all do the same.

Basic landing-gear form shaped from 2x4 and covered with clear MonoKote.

Landing gear on form ready to be vacuum-bagged.

78 WORKSHOP SECRETS

LANDING-GEAR ESSENTIALS

TOOLS AND MATERIALS YOU'LL NEED

- 3-ounce, unidirectional carbon-fiber cloth.
- West Systems or EZ-Lam epoxy.
- Epoxy mixing cups and application sticks.
- A disposable 1-inch brush with the bristles trimmed down to ½ or ¾ inch.
- A pair of "shop" scissors to cut the carbon fiber.
- A scrap piece of pine 2x4, at least 8 inches long.
- MonoKote.
- Scrap balsa wood.
- 2-56, socket-head screws, nuts and washers.
- Wheels.
- A drill with a ¼-inch bit.
- Disk sander or sandpaper on sanding block.

To vacuum-bag your parts:
- A vacuum pump.
- Peel-Ply.
- Bleeder material (paper towels will work for small projects like this).
- Sealing film or a gallon-size, sealable sandwich bag.
- Window-caulking putty.

fibers are saturated, any additional resin only adds weight and reduces strength. Vacuum-bagging helps to remove any remaining excess resin as well. When the top layer is saturated, apply the cross-grain fibers at the axle mounts and the center. If you are not planning to vacuum-bag the gear, set the mold aside and let the resin cure overnight.

without wax on the MonoKote, the finished product was released with little trouble.

You should prepare everything before you mix your epoxy. Clean up your workspace and cover the bench with cardboard or plastic. Measure the total length of the landing gear going around the curved surface of the mold, add a little length to be safe, and then cut all the carbon fiber that you'll need for the layup. Decide how much material it will take for the project, either by analysis or by trial and error. After you've made some carbon-fiber parts, you will be able to estimate what you will need without doing either. I decided on three layers of cloth cut into 1-inch strips. While cutting the cloth, also cut a few narrow ½- to ¾-inch-long scraps that you'll apply "cross-grain" near the axles and at the center; this will increase the strength of these high-stress areas. Be careful: any loose particles of carbon fiber will float around the workshop and create a hazardous environment. A respirator and a fan are very good ideas.

When all the carbon has been cut, dispense one pump each of resin and hardener into a plastic yogurt cup. Mix them thoroughly for about a minute.

Work with just one layer at a time when you lay up the landing gear. First, place the bottom layer of carbon on the curved surface of the mold and apply some epoxy. With the disposable brush, carefully work the resin into the fibers using a stippling or poking motion. Eliminate any air bubbles or voids. Be careful to keep the fibers straight, and fully saturate them with resin. It is OK to use a little excess epoxy at this point because it can be worked into the upper layers as you go.

Now place the next layer of carbon on the form, and work any excess resin from the first layer up into it. If necessary, add more resin to fully saturate the new layer. Continue to add your layers in this manner, gently stippling the resin thoroughly into the fibers. Reduce the amount of resin you add as you get near the end of the layup process; when the

VACUUM-BAGGING

It is best to vacuum-bag the landing gear to achieve the optimum resin-to-fiber ratio. In addition to the vacuum pump, you will also need Peel-Ply, bleeder material, some sort of plastic film and window-caulking putty. Peel-Ply is a Teflon-coated nylon fabric that allows excess resin to migrate out of the layup and into the bleeder material. In addition to absorbing excess resin, the bleeder material helps get all the air out of the sealed system. The air can easily flow in and around the fibers of this material and escape. Without the bleeder, one area of the mold can be sealed off from the rest, and this will

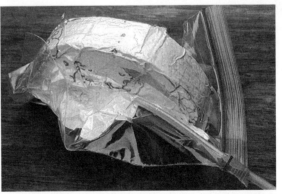

The complete layup in a vacuum bag. Notice that the vacuum tube is against the mold and that the mold and the tube are surrounded by paper-towel bleeder.

prevent the whole mold from being in a vacuum. For small parts like this landing gear, you can just use a couple of layers of paper towels for the bleeder material. To seal a small, hard mold like this, it is easiest to use a plastic bag; in this case, a gallon-size Ziploc freezer bag worked well.

The actual vacuum-bagging process is quite simple. When you finish the layup, place a layer of Peel-Ply completely over the new part. Cover everything with the bleeder material, then slide all of this into the freezer bag. Before you close the bag, insert the hose from the vacuum pump, placing it against the side of the block. Make sure that there is some bleeder material around the end of the hose so that the bag will not get in the way. Close the bag's "zipper," and with some window-caulking putty, seal around the vacuum hose. Turn on the pump and seal any leaks. Pull as much vacuum as you can—ideally, 28 to 29 inches of mercury. Let it cure overnight.

Completed landing gear with wheels. Harden the balsa locator block with thin CA.

FINISHING UP

When your gear has cured, unwrap it, and pop it off the mold. Some cleanup will be necessary before the part is ready for use. I used a bench-mounted disc sander for this and a shop vac to control the carbon dust. You can use sandpaper mounted on a sanding block, but then you won't be able to control the dust. Be very careful! The dust created by sanding, grinding and cutting carbon fiber is a health hazard. Do not breathe any of it. I use a closely fitting canister respirator and work outside in a breeze to avoid inhaling the dust.

After cleaning up the carbon fiber, drill the ends of the gear for axles. I used 2-56 socket-head cap-screws, nuts and washers, and they were strong enough for a model as light as the Dragonfly.

Mounting the gear on the model was simple: I drilled two small holes in the top of each gear about ¼ inch from the center. After threading a small plastic wire tie through this hole, I cinched it around the center carbon tube. Note that the photos show the cross-grain carbon fiber at the center of the landing gear and where the axles are. This prevents the wire tie from splitting the gear parallel to the fibers. I shaped a block of balsa to fit between the sides of the plane and to cradle the tube fuselage. Medium CA will harden the blocks and will also attach them to your gear. The blocks restrict the side-to-side motion of the landing gear, and they also transfer some of the shock of landing to the center tube.

CONCLUSIONS

My experiment was a complete success. Flight-testing determined that the CG shift was not significant for this model. The new gear even caused less drag than the original gear. The flat plate of carbon is less than 1/32 inch thick and has a much more aerodynamic cross-section than the round wire. Ground handling was greatly improved as well. The wheels maintain better alignment, and the wings are more likely to stay level in a crosswind and while turning. Harder landings just flex the gear a bit, but the plane doesn't bounce back up into the air. Repeated landings and touch-and-go's have proven this landing gear to be much more durable and sturdy than the original wire gear. Without my really trying to save weight, the landing gear weighs only 6.4 grams—just 0.2 gram more than the original music-wire gear. Another few seconds at the disc sander would easily remove that, but I'm having too much fun to take the model apart again!

What about larger and smaller models? The beauty of these materials and techniques is that you can scale the layups to accommodate any size of model. While the Dragonfly is the lightest model for which I personally have made carbon-fiber gear, you could easily make parts for smaller and lighter models merely by reducing the amount of carbon and epoxy used and resizing the mold to match your model. On the upper end of the scale, there are full-scale aircraft. A few years ago, I spent a number of hours fitting retractable gear to a homebuilt canard. Very similar to the famous Long-Eze, it, too, uses solid carbon and epoxy gear legs. Anything's possible with carbon fiber.

The underside of the landing gear that has been mounted to the fuselage. A small scrap of carbon was glued cross-grain between the holes to prevent splitting.

The top of the now mounted landing gear. Note wire tie and balsa locater block.

3
Control-linkage setups

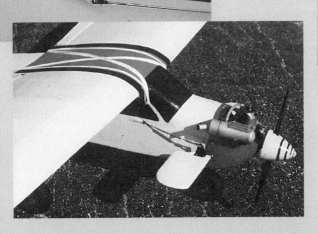
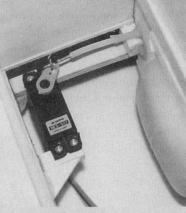

CHAPTER 3

Easy Z-bends

by Randy Randolph

Modelers became acquainted with Z-bends about the time of the first U-control models. Up until then, there was little application for that funny-looking "quirk" in the end of a piece of wire as far as model airplanes were concerned. They have since become a way of life as well as a frustration when it is necessary to form one in the end of a piece of steel wire. The photos show a smooth and easy way to form these fancy bends in wire of almost any size.

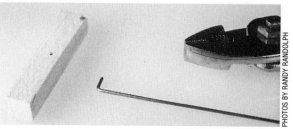

3 Use the vise or a pair of pliers to make a right-angle bend in the end of the wire. This will form the latch of the completed bend and should be nearly twice as long as desired (¼ inch for an ⅛-inch latch).

4 Insert the latch bend into the hole in the hardwood, and clamp the wire and wooden block tightly into the vise. To provide the hook width desired (here, ⅛ inch), the top of the wooden block should be even with the top edge of the vise jaws. The wire should be perpendicular to the vise and wooden block.

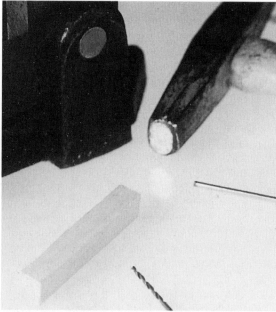

1 The necessary tools include: a vise, a hammer, a piece of hardwood and a drill bit that's the same size as the wire to be bent. The harder the hardwood, the better, but pine will work for wire with a diameter that's smaller than 1/32 inch.

5 Hold the other end of the wire and bend it straight back from the jaws while tapping with the hammer at the site of the bend. Let the hammer do the work. When the wire is flat against the vise jaws, the bend is complete, and the wire can be removed from the vise.

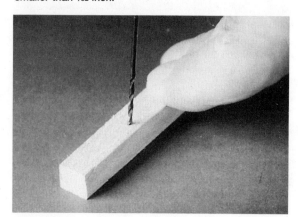

2 Since this bend will be in 1/16-inch-diameter wire with a ⅛-inch hook and latch, drill the 1/16-inch hole ⅛ inch from the edge of the hardwood block. Drill the hole at least twice as deep as the intended latch (in this case, ⅛ inch).

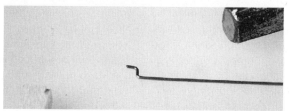

6 The latch can now be trimmed to the desired length (⅛ inch in this example). Z-bends in larger wire can be made in this way if a mild steel (rather than hardwood block) is used.

CONTROL-LINKAGE SETUPS

Make sewn hinges
by Bob Underwood

The increased interest in slow-flight, indoor RC and small RC airplanes has made it necessary to hinge control surfaces in a way that offers the least resistance to miniature and sub-miniature servos, as well as magnetic actuators. The hinge that offers the least resistance to movement is one that's sewn with a baseball stitch and silk thread. It's probably the strongest, lightest hinging system available to modelers. The photos show the way.

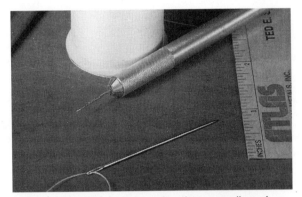

1 The only tools necessary, other than a needle and thread, are a ruler and a pin. The pin is much easier to use if you remove the head and insert it in a pin vise or modeling knife. Both surfaces to be hinged should already be covered.

2 The job will be much easier if you prepunch the needle holes with the pin. Using the ruler as a gauge, punch the holes through the mating surfaces at the hinge locations. One-eighth-inch centers are ideal, and about five or six holes in each surface are usually sufficient.

3 Thread the needle with a 12-inch loop of thread and tie a knot in the end of the loop. Start with the first prepunched hole and insert the needle from the bottom of one surface, pulling the thread through until the knot touches the surface.

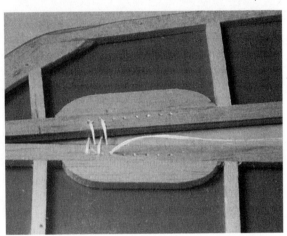

4 Bring the needle between the surfaces and through the bottom of the mating surface, then again between the surfaces and through the bottom of the first surface. Notice that the stitch always starts at the bottom side of each surface.

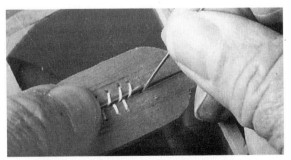

5 After two or three stitches, pull the thread up tight and hold the two surfaces in place while you finish sewing them together. The needle will pass easily between the surfaces, and after the last stitch, gently pull the thread to remove any slack.

CHAPTER 3

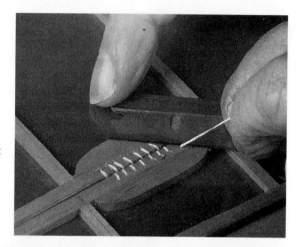

6 A drop of glue at the knot and at the end of the last stitch will hold them permanently in place. When the glue has set, trim the thread and the knot flush with the surface.

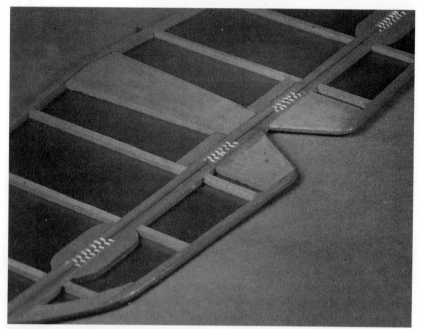

7 The finished hinge line is tight and very free-moving. In this case, four sets of hinges do an adequate job on the elevator. Two hinges will usually suffice for the rudder. Although it looks clear, this stab is covered with Oracover Light Transparent film. The complete hinge job added less than a gram to the total weight of the stab.

CONTROL-LINKAGE SETUPS

Install direct-control aileron servos
by Gerry Yarrish

One of the most popular modifications modelers make to their planes is to replace the long wire pushrods and 90-degree bellcranks that handle aileron control with servos mounted directly in front of the control surface. This eliminates many linkage contact points and goes a long way toward reducing control-system slop. Here's one way to do it.

1 This is a typical, built-up wooden wing in which a bellcrank and a long music-wire pushrod would usually be installed for aileron control. In this case, I'm modifying a Model Tech P-51 ARC Mustang wing. The open-bay wing construction makes this a very easy conversion, but if your wing has capstrips installed on the ribs, you'll need to remove them before you begin. For a fully sheeted wing, first modify the rib, and install the servo before the wing is sheeted. Also, a fully sheeted wing will need a flush-fitting hatch cover for servo access.

2 First, install a cardboard tube in the wing (you'll route your servo wire through this tube). I used tubes from the Aeroplane Works. They slipped easily into the precut rib openings, and I glued them into place with thick CA. To cut holes in your ribs, use a piece of sharpened brass tube, and start from the root rib and work out toward the rib bay where the aileron servo will be installed.

4 To supply more material for the servo screws to "bite" into, add two 3/32-inch-thick strips of plywood to the other side of the rib. After the strips have been glued into place, drill into them through the holes in the mount plate. Use a 1/16-inch-diameter drill bit for standard-size servo screws.

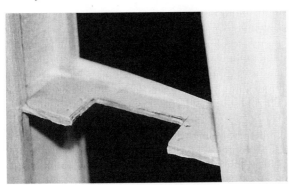

3 Next, cut out a piece of 3/32-inch-thick plywood (not lite-ply) to fit your servo and the rib it will be attached to. Before I glued the plywood to the rib, I cut the servo opening in the plywood mount plate and drilled holes for the servo screws. After the ply has been glued into place, cut out the exposed part of the rib where the servo will go.

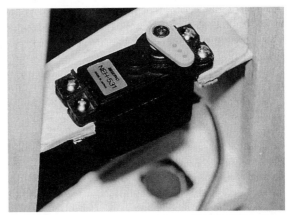

5 Place your servo (here, I'm using a JR NES-531) in the cut-out area of the mount plate, and screw it into place with the screws that come with it. For good servo support, make sure you install the rubber grommets and the brass inserts properly. When the servo has been installed in the wing, it should be flush with the bottom of the rib.

6 The next step is to glue a 1½- to 2-inch-wide capstrip to the rib that supports the aileron servo. The capstrip must be wide enough to allow you to cut a ¼-inch-wide slot into it for the servo arm to pass through and to accommodate the width of the pushrod clevis that will be installed on the arm later. I made my slot 2 inches long.

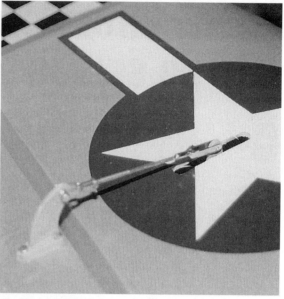

9 The completed aileron control linkage setup. This conversion is very easy to do and doesn't take much time. If, at a later time, you need to remove the servo or inspect it, simply cut away a portion of the covering between the ribs. A patch of material can then be ironed back into place and will be nearly invisible. Enjoy!

7 Here's the finished installation. Notice the servo arm protruding through the slot. To sand the capstrip flush with the wing's leading- and trailing-edge sheeting, simply rotate the servo arm until it's beneath the capstrip, and sand it using a sanding block.

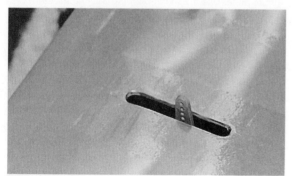

8 After the wing has been covered, use an X-Acto knife to cut away the covering over the slot. Then rotate the servo arm until it's in the neutral position, and make up and install your aileron control pushrod.

CONTROL-LINKAGE SETUPS

Install Robart hinges in ARC models
by Gerry Yarrish

Like many modelers, when I build an aircraft, I have particular hardware preferences. I choose Robart HingePoint hinges because they are very easy to install and, I think, look more true-to-scale than the standard "flat" hinges.

But if you build an almost-ready-to-cover (ARC) kit, you'll have to make a few internal modifications if you want to install HingePoint hinges properly. Here's my quick fix for the ARC hinging problem.

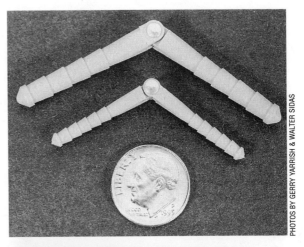

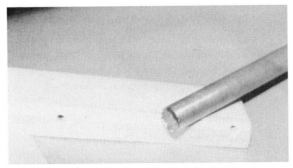

1 Mark where the hinges will be installed. Inside the control surface, you need enough solid balsa to make the ¾- to 1-inch-deep hole that you'll drill for each hinge. I install balsa blocks; for each block, I cut an access hole in the control surface's sheeted skin with a sharpened, large-diameter brass tube.

2 To cut the access hole, gently twist the tube while applying firm pressure. Cut all the way through the balsa, remove the neat little balsa plug from the end of the tube, and save it to close the hole after you've installed the block.

3 Here are a plug and a balsa block ready to install. Carve the block a little to make it fit the hole: knock off the corners and carve it into a wedge shape.

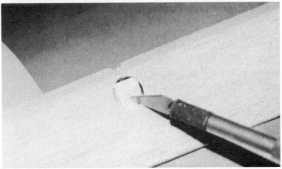

4 Put the block into the hole, then push it up against the leading edge with a hobby knife, and secure it with thin CA.

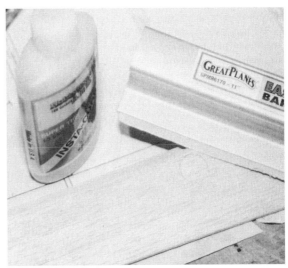

5 Put the balsa plug into the hole, and glue it into place with CA. Here, I used Bob Smith's Insta-Cure Super Thin. When this has cured, sand the plug flush with the control surface's outer skin. (A Great Planes Easy Touch Bar Sander is useful here; I use 220-grit sandpaper.)

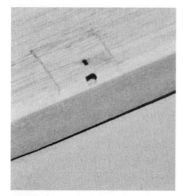
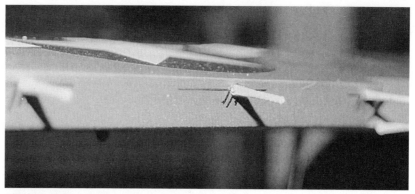

6 If you need to install larger HingePoints in a giant-scale model, simply use a hobby knife to cut a larger, square opening in the control surface, as shown here. Install a balsa block that's as large as necessary to properly distribute the hinge load.

7 With all the balsa blocks glued into place and all the holes closed, it's time to cover the model. To position the hinges properly, draw horizontal and vertical centerlines for each one (I use a Sharpie marking pen). Then drill the holes and fit the hinges temporarily in the wing trailing edge.

8 Next, butt the control surface (here, an aileron) against the wing trailing edge, and transfer the hole marks to the control surface's leading edge so that the holes in the wing and the control surface match. Drill the holes, and temporarily fit the hinges in the control surface. Here's a tip: to minimize the gap between the control-surface leading edge and the wing trailing edge, use a rat-tail file to notch the control-surface leading edge at each hinge location. This allows you to push the hinges farther into the leading edge while allowing them full motion.

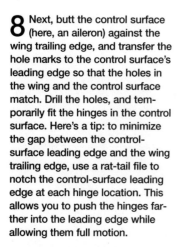

9 Finally, trial-fit the hinges, and when you're satisfied with their fit, glue them all into place with an alphatic-type glue or Pacer Technology's new Hinge Glue. Before you install the hinges, be sure to completely coat the inside of the holes with glue. Periodically check the hinges for freedom of movement, and when the glue is dry, install the control horns and linkage.

So, even though you build ARCs, you can still use Robart HingePoint hinges. The balsa blocks greatly increase the strength of the installations, and remember: if all else fails, improvise!

Control-linkage setup
by Gerry Yarrish

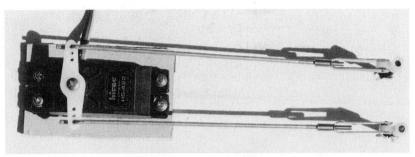

A typical aileron linkage setup for a .40-size sport model. The clevises are attached to the vertical arms of the aileron torque rods but no stop nuts are installed. Stop nuts are very good insurance against slop and should be used whenever possible.

Whether you're building your first high-wing trainer or scratch-building an original-design, giant-scale unlimited racer, you have to connect servo movement to control surfaces. There are many types of setups, and hobby shops are full of hardware to accomplish this task. Here's some basic setup information that you can apply to your model.

LOW-POWER TRAINERS

In a typical trainer (.25- to .45- size), the beginner can do no better than follow the manufacturer's directions. In most cases, elevator and rudder are actuated by solid-wood pushrods. The pushrods typically have metal wire ends so that the pushrod can be connected to the servo (with a Z-bend) and to the control surface (by means of a threaded end and a clevis). Figure 1 shows the most basic setup for elevator and rudder. Usually, ¼- to ⁵⁄₁₆-inch-square balsa stick stock is used for the pushrods. The 2-56 pushrod wires are bent 90 degrees as shown and inserted into a hole drilled through the pushrod. The wire rod end is then bound with thread and glued (some trainers use shrink-tubing in place of thread).

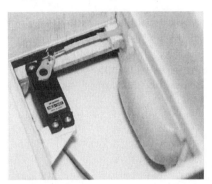

Quick-type links, such as this one on my throttle servo, are good for easy adjustment and simple setup. Don't use these links on your elevator servo.

Ball links used in this throttle-linkage setup minimize slop and allow for slight misalignments while the throttle is advanced and retarded.

SPORT MODELS

Once past the trainer stage, most modelers move up to faster, more powerful sport models in the .40- to .60-size range. Low-wing and other high-performance models generally include aileron control as part of the linkage installation.

Harder wood, such as birch dowels, is needed for these types of models. These dowels work well and

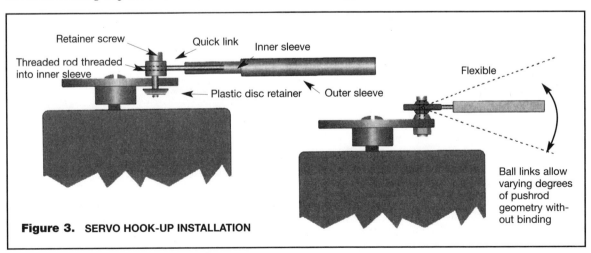

Figure 3. SERVO HOOK-UP INSTALLATION

CHAPTER 3

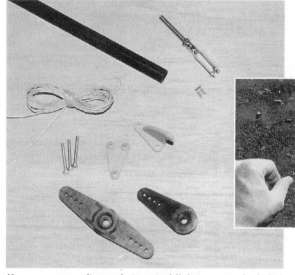

Here are some aftermarket control-linkage parts. Included are a carbon-fiber, arrow-shaft pushrod; a 4-40 threaded rod and clevis (notice the clip retainer); Kevlar pull/pull thread; Robart's ball-link control horn; and Du-Bro reinforced giant-size servo arms. These extras can increase your model's longevity.

To save weight, many modelers also use hollow carbon-fiber or fiberglass arrow shafts for pushrods. Larger diameter (3/8 inch) pushrods are even used in some giant-scale unlimited racers.

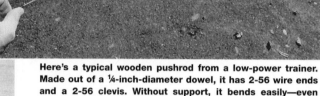

Here's a typical wooden pushrod from a low-power trainer. Made out of a 1/4-inch-diameter dowel, it has 2-56 wire ends and a 2-56 clevis. Without support, it bends easily—even with slight finger pressure.

provide stiff control response; however, they need to be kept as short as is practical to minimize vibration and possible flutter. Supporting the middle of the pushrods with a bulkhead or a couple of cross-braces will do the job nicely, but the friction between the rods and the support should be minimal.

The flexible or "tube in a tube" pushrod is also popular and works well if it's installed properly. The outer sleeve must be supported every 6 to 8 inches along its length, and the inner sleeve must slip easily through the outer sleeve without binding. Bends in the pushrod must be gradual and smooth.

PULL/PULL CONTROL

Typically, a pull/pull control setup consists of thin, braided, metal cables or Kevlar thread. Pull/pull

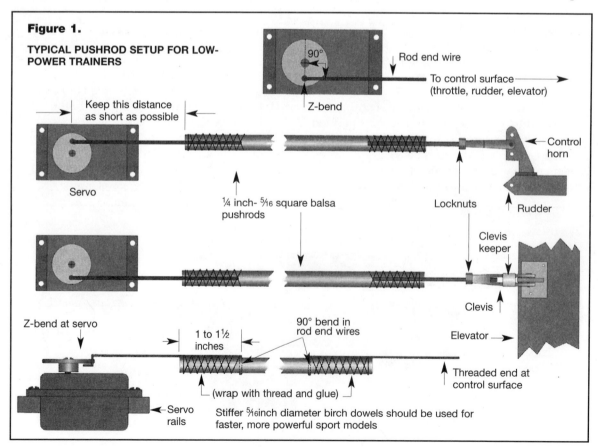

Figure 1. TYPICAL PUSHROD SETUP FOR LOW-POWER TRAINERS

CONTROL-LINKAGE SETUPS

control is very popular among scale modelers and pattern fliers who demand precise rudder control. Figure 2 shows the ideal setup.

PUSHROD HARDWARE

The size of control horns, clevises and threaded rods depends on the size of the model. The 2-56 hardware is good for small models (up to .40-size), and it can also be used in larger, less powerful models, such as .60-size Cubs. The larger, stiffer 4-40 hardware is well-suited to giant-scale hot rods, such as Lasers and Extras.

BALL LINKS

Ball-link clevises and control horns provide precise control action and allow varying degrees of pushrod geometry without drag or friction. Ball links should be used only at the servo end of a pushrod and not at the control horn. Placing an "offset" ball link at the control horn causes an out-of-alignment connection that tends to twist the horn under severe flight loads. Robart's new ball-link control horns are the exception because the ball link is captured "in line" with the control-horn's arm.

A good place for ball links is at the engine's throttle-control arm. In gas engines, where there is usually a bellcrank, ball links automatically compensate for misalignment, and they minimize slop.

QUICK LINKS

Quick links are small metal attachments that are held to the servo's arm with a plastic disc, and they have holes drilled in their sides that accept a cable or a music-wire rod end. These are good for simple installations, such as the throttle, but they shouldn't be used for elevator control. Again, the size of the model and the flight loads will determine which type of connector to use.

SERVO ARMS

Most models work very well with the standard arms that come with the servo. There are now a number of aftermarket servo-control arms that are bigger, stiffer and stronger than the stock ones. Generally speaking, the pushrod should always be 90 degrees to the control arm in the neutral position. An exception to this is in aileron setups where you may want differential control, but this will be discussed in a future control-linkage setup article. Figure 3 shows some typical servo installations.

For proper control of your model, you have to know the basics of control-linkage setup. Minimizing slop and maximizing control stiffness will help keep your model in one piece.

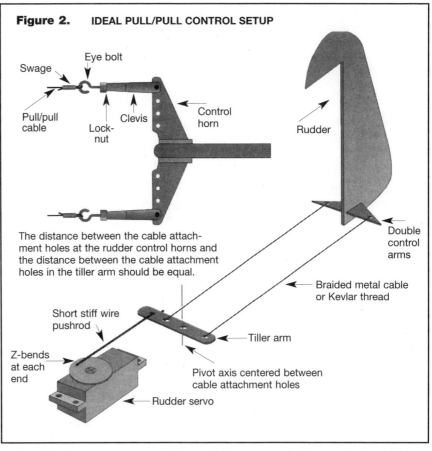

Figure 2. IDEAL PULL/PULL CONTROL SETUP

The distance between the cable attachment holes at the rudder control horns and the distance between the cable attachment holes in the tiller arm should be equal.

CHAPTER 3

Control linkages   by Mike McConville

In many models, the ailerons and the flaps can not be connected to the servo with a simple torque rod; they must be connected either to a single servo via bellcranks, or to dual servos mounted outboard in the wing. Often, your choice of method depends on which model you are building, because some models don't have room in the wing for two servos. A bellcrank system, on the other hand, is usually built into a wing; because you can't get to this system to maintain it, install it with durable hardware.

DUAL SERVOS OR BELLCRANKS?

For aileron and flap linkages, I prefer to use outboard wing-mounted, directly connected dual servos. For a small additional cost, you'll have a rigid servo connection that is easy to install and maintain. In higher-powered performance models, this rigidity adds a good margin of safety against flutter.

If wing-mounted servos are not a possibility, install bellcranks. Because there are so many pushrod connections and moving parts in a bellcrank system, it's easy for "slop" to develop. Use high-quality hardware for the linkage connections from the servo to the bellcrank and from the bellcrank to the control horns. Because these linkages are built into the wing, any repairs will require major surgery. Connecting the linkage from the bellcranks to the servo can pose a problem, but the Du-Bro aileron connector ball link (part no. 183) is made specifically to remedy this; it works well.

SETUPS

• **Bellcrank orientation.** For aileron linkages, orient the bellcranks in opposite directions; this will produce the desired aileron movement. For flap linkages, orient both bellcranks in the same direction, as shown in Figure 1; this will produce movement in the same direction.

• **Servo-arm orientation.** To achieve the correct movement when you use dual servos and a "Y" harness to plug both servos into the same receiver channel, the aileron servo arms should be oriented in opposing directions, and for flaps, they should both be on the same side as shown in Figure 2.

AILERON DIFFERENTIAL

Often, you'll want differential in the aileron movement. "Differential" here simply means more movement of the control surface "up" than "down." Use aileron differential to correct adverse

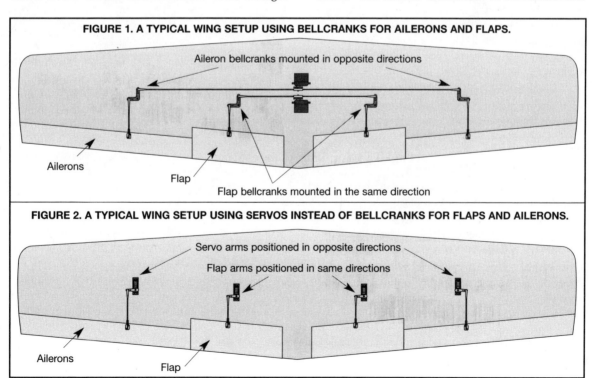

FIGURE 1. A TYPICAL WING SETUP USING BELLCRANKS FOR AILERONS AND FLAPS.

FIGURE 2. A TYPICAL WING SETUP USING SERVOS INSTEAD OF BELLCRANKS FOR FLAPS AND AILERONS.

CONTROL-LINKAGE SETUPS

yaw and to control the roll axis of an aerobatic model.

• **Differential via bellcrank position.** Figure 3A shows a bellcrank system with no differential movement. In this setup, the arm of the bellcrank that is connected to the aileron is perpendicular to the pushrod, and the arm that is connected to the servo is perpendicular to its pushrod when at neutral. Figure 3B shows the neutral position of a setup that will produce more "up" than "down" movement, assuming that the control horns are on the bottom of the control surface. Figure 3C shows the setup that will produce more "down" than "up" movement, assuming the same horn attachment. The drawback to this method of setting differential is that, once the model has been completed, it is not easily adjusted.

• **Differential via splined servo-output shaft.** You can adjust differential by changing the orientation of the servo arm by rotating it to a different position on the splined servo-output shaft. Again assuming that the horn is on the bottom of the control surface, Figure 4A shows the servo arm position with no differential, 4B shows how to produce more "up" than "down," and 4C shows how to produce more "down" than "up." The farther the arm is rotated from the zero position, the greater the differential movement. This method is preferred because it is usually easy to get to the servo so you can make adjustments quickly while you are trimming a plane.

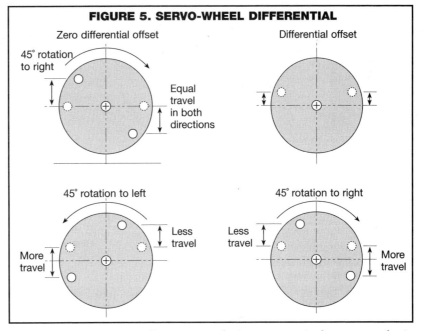

Aileron differential with bellcranks. All control horns are mounted on the bottoms of the ailerons.

Figure 3A. This setup will result in zero differential.

Figure 3B. This setup will result in more "up" than "down" aileron movement.

Figure 3C. This setup will result in more "down" than "up" aileron movement.

Aileron differential with servos. All control horns are mounted on the bottoms of the ailerons.

Figure 4A. This setup will result in zero differential.

Figure 4B. This setup will result in more "up" than "down" aileron movement.

Figure 4C. This setup will result in more "down" than "up" aileron movement.

FIGURE 5. SERVO-WHEEL DIFFERENTIAL

Zero differential offset — 45° rotation to right — Equal travel in both directions

Differential offset

45° rotation to left — Less travel / More travel

45° rotation to right — Less travel / More travel

To set up and trim out an airplane properly, it is important to have a good, practical understanding of the mechanics of control linkages.

CHAPTER 3

Control linkages for giant-scale models
by Mike McConville

Bigger is better. When it comes to today's model airplanes, there's certainly a strong argument for this statement. Giant-scale airplanes are becoming increasingly popular, but "the world of the giants" presents many factors for consideration. Control-linkage requirements are becoming ever more critical to safety and good performance.

MEASURING UP TO THE TASK
Control systems on typical giant-scale airplanes—1.20 size to 5.8ci and bigger—require more attention than those on smaller models; the bigger the model, the greater the dynamic loads on the airframe and control surfaces. In a smaller model, the integrity of the linkage systems often far exceeds the loads that they see. But on giant-scale models, the linkages are pushed to their limits. There are several areas in control systems that require special attention, and this month, we'll focus on three parts of control linkages that must be correct.

• **Linkage rigidity.** All control linkages must be rigid and free of flexing, or the control surface will not track directly with the servo; at best, the model will feel "squirrely," and at worst, the linkage will fail, and the model will be lost. In most cases, 4-40 hardware is preferable (particularly on bigger gas-burning models), though this depends on the size and type of model. To achieve short, rigid linkages,

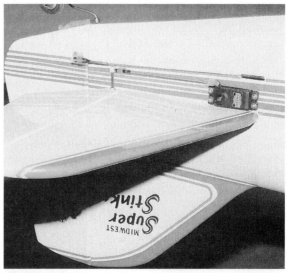

The tail-mounted elevator servos in my Super Stinker keep linkages short and very rigid. I used Hayes clevises at the horn attachment and soldered clevises at the servo-arm attachment.

tail-mounted servos are often used; I used them in the Midwest Super Stinker.

For some models, tail-mounted servos do not work because they make the plane tail-heavy or spoil the scale look. In such cases, very stiff pushrods (such as carbon-fiber—available from Aerospace Composites, or fiberglass—available from Dave Brown Products) should be used. As shown in the illustration, the pushrod should be supported at the rear to prevent side-to-side movement. Use a short length of brass tube supported by a brace as a bearing for the pushrod to slide through.

• **Mechanical advantage.** When setting up and adjusting control-surface throws on a small model, we don't usually consider this factor. When we want more throw, we move the linkage attachment point on the control horn closer to the control surface. This shortens the lever arm and decreases the linkage's mechanical advantage, i.e., leverage over the surface. On smaller planes, the forces are usually so small that we will not run into trouble.

On giant-scale models, particularly on high-performance aerobatic types that have large control surfaces, this is an important concern; it isn't safe to move the linkage attachment point closer to the surface to achieve more throw. The loss of mechanical advantage may push the control system past a critical point, and severe flutter may result. Always keep the control horn or lever arm as long as is practical.

If more throw is needed, install a longer arm on the

On the stab of my TOC Extra 300S, I used long control horns to maintain mechanical advantage and long, heavy-duty SonicTronics servo arms to achieve the necessary control throw; 4-40 hardware is used throughout.

CONTROL-LINKAGE SETUPS

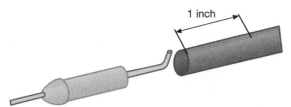

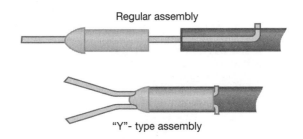

Hollow fiberglass pushrods (available from Dave Brown Products) provide rigid, light transfer of servo power to the control surface. Follow the manufacturer's instructions for proper installation.

servo, and move the linkage attachment point farther away from the servo-output shaft. Long, glass-filled, nylon, heavy-duty arms available from Du-Bro and SonicTronics or aluminum arms available from Hangar 9 are good choices. The photo on the preceding page shows this principle on the elevator

Du-Bro's strong, long, fiber-filled, heavy-duty servo arms are good for increasing control throws and come with spline patterns for all the major radio brands.

• **Linkage attachments.** These must be secure and free of binding, slop and "twist." Ball-type links are usually preferable because they have no slop, they wear very well and are very secure. The control horn must be rigid and must not flex. The photo of my TOC Extra stab shows Rocket City giant-scale control horns, which work very well. They consist of an 8-32 bolt (which serves as the control horn) and ball-attachment links. Clevis attachment to the control horn works well, but take care to choose a clevis and horn that don't have slop between the clevis pin and the hole in the horn. The clevis and horn method requires periodic inspection, because the hole in the horn will become worn and the pin's fit will be sloppy. Whenever slop develops, the hardware should be replaced. Avoid metal clevises that are sloppy on the threads of the pushrod; this slop will

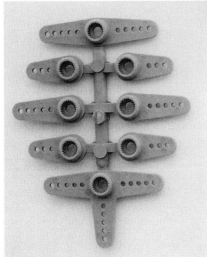

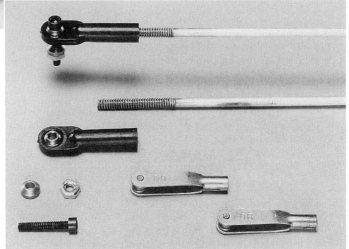

Du-Bro 4-40 metal clevises and ball swivels should be used on giant-scale models; 4-40 hardware is stronger than 2-56 hardware and better suited to the surface loads that giant-scale models endure.

linkages of my 37-percent Extra 300S, which I flew in the 1994 Tournament of Champions. While practicing for the last TOC, I saw an Extra exactly like mine, using the same elevator servos and powerplant, have elevator flutter in level flight at half throttle; the only difference was its very short elevator horns, which were intended to achieve radical elevator throws.

hurt the model's performance. As you can see in the photo of the Super Stinker tail, Hayes Products makes a high-quality plastic clevis with a metal pin; it can be threaded onto the push-rods tightly, and the metal pin will not become worn. Attachment to the servo is

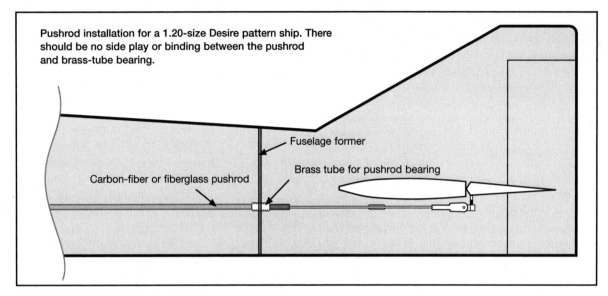

Pushrod installation for a 1.20-size Desire pattern ship. There should be no side play or binding between the pushrod and brass-tube bearing.

much the same. I prefer Du-Bro swivel links—a ball-type attachment. Z-bends work fine on small sport models, but don't use them on large models. They are usually somewhat sloppy, and they'll wear the hole in the servo arm.

When you think about it, flying an airplane that has sloppy or flexible linkages yields the same effect as flying one that has servos loose in their mounts. With "big birds," there's the added danger of failure caused by an incorrect linkage.

CONTROL-LINKAGE SETUPS

Stop control-surface flutter
by Mike McConville

Flutter in your rudder—or in any control surface for that matter—usually produces disastrous results. There are many factors that can cause flutter. This month, I'll look at control-system setup and examine the most common causes of, and practical remedies for, control-surface flutter.

WHAT'S FLUTTER?
Flutter is the unwanted oscillation of a control surface in flight. Depending on its severity, flutter can range from a barely audible hum to a trailing edge that looks like a complete blur. The outcome always results in some type of failure—a broken linkage, stripped servo gears, failure of hinges or of the structure to which the fluttering control surface is attached. The result is a seriously damaged airplane. The causes of flutter can be very subtle and easily overlooked. How many times have you seen two seemingly identical airplanes, but one worked well and the other had a flutter problem?

AN OUNCE OF PREVENTION
Before flying any new airplane, seal the control-surface hinge gaps. Often, the hinging seems to leave a gapless connection, but no unsealed gap is completely airtight. Always seal the gap, regardless of how tight it may seem. This reduces the possibility of flutter and results in more positive and effective control of the model. (I consider this step a normal part of building.) Gaps can be easily sealed from the bottom of the surface with a strip of iron-on covering (see Figure 1).

CONTROL-SURFACE TWIST
The twist might have been good for Chubby Checker and might even be a good tune for flying a freestyle routine, but it isn't good for a control surface. Twist or torsional flex in a control surface often occurs in a long, solid, balsa surface, such as a strip aileron. If the balsa is too soft, the surface may easily flex and act as if it's disconnected from its linkage. You should replace extremely soft, flexible balsa control surfaces. I've seen this happen many times on Quickie 500 racers because of their high speeds and very light airframes. A very easy fix is to partially tape (use a flexible tape, such as 3M clear decorator tape) the underside of the aileron tip to the fixed portion of the wing (see Figure 2).

Du-Bro 4-40 metal clevises and ball swivels should be used on giant-scale models; 4-40 hardware is stronger than 2-56 hardware and better suited to the surface loads that giant-scale models endure.

This makes the surface twist somewhat when it's deflected, but it usually doesn't result in a loss of control rate. This method also works well on Sunday fliers.

MASS BALANCES
The need for mass balancing is much greater on larger models. The bigger a control surface is, the more mass it has. As the surface mass increases, so does its inertia. In practical terms, this means the heavier the control surface is, the more its mass will contribute to the surface oscillation. After a point, the linkage system can not overcome the surface load, and flutter will occur. I've often heard people say that they had beefed up the control surfaces to stiffen them but they still fluttered. In reality, beefing up the control surfaces increases the surface mass, which causes flutter. Mass or counter balances are simply the addition of weight to the control surface ahead of the hinge line so that the surface balances level or at least close to level when

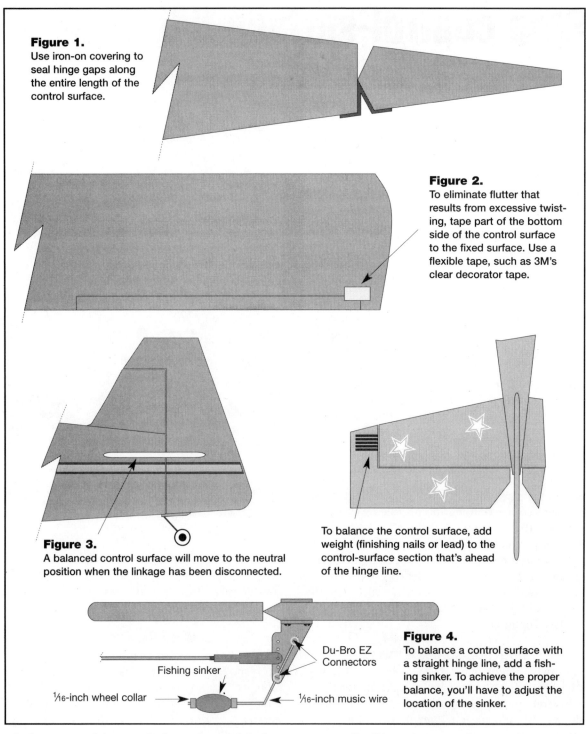

Figure 1. Use iron-on covering to seal hinge gaps along the entire length of the control surface.

Figure 2. To eliminate flutter that results from excessive twisting, tape part of the bottom side of the control surface to the fixed surface. Use a flexible tape, such as 3M's clear decorator tape.

Figure 3. A balanced control surface will move to the neutral position when the linkage has been disconnected.

To balance the control surface, add weight (finishing nails or lead) to the control-surface section that's ahead of the hinge line.

Figure 4. To balance a control surface with a straight hinge line, add a fishing sinker. To achieve the proper balance, you'll have to adjust the location of the sinker.

Du-Bro EZ Connectors
Fishing sinker
1/16-inch wheel collar
1/16-inch music wire

it's disconnected from its linkage. On models that have counter-balances as part of the control surface, balancing can be achieved by adding weight to the control surface ahead of the hinge line. Figure 3 shows this type of setup on the elevators of the new Midwest Giles G-202. On models that have a straight hinge line, a balance can be added to the bottom of the control surface. As shown in Figure 4, a balance can be added to the control horn. If the flutter is not caused by unsealed hinge gaps, a flexible twisting surface or an improper linkage setup, mass balancing will almost always cure the problem.

In conclusion, if flutter is detected, land immediately, and take measures to correct the problem. Flutter's destructive force can literally tear an airframe apart; we all put too much time and effort into our models to take unnecessary risks.

CONTROL-LINKAGE SETUPS

Commonsense control linkages
by Greg Hahn

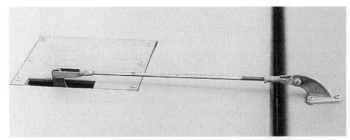

Good aileron linkage: the servo is attached to a plywood plate, and the servo arm extends through the surface. It's connected to the aileron with a 4-40 rod and a Robart horn. This short, sweet setup has good leverage, and it's easy to keep an eye on it.

When leafing through my notes of frequently asked questions, I came across one that I get hit with all the time and thought that it might make a good subject to write about. Of course, I always seem to find controversial subjects, and—as usual—my way of doing things isn't always the norm!

The issue in question concerns servos and linkages: how you go about operating the many surfaces and control functions of today's giant-scale aircraft. As far as I know, there aren't any published rules of thumb on just what to do and how to do it, so I'll throw my standard setup opinions, equipment and procedures into the fiery depths of public scrutiny and maybe come up with some good rules to go by.

In the late '70s, Nick Ziroli and a handful of other visionaries pioneered the idea that model aviation could involve more than small "toy" airplanes. Now, giant-scale airplanes, contrary to their humble beginnings, are adorned with everything that technology has to offer, from computer radios and onboard telemetry to high-horsepower, finely tuned gasoline engines. Yet even with all this techno stuff to the forefront, the all-important control linkages are still in question, and most of the available hardware is still a scaled-up version of a standard 40 to 60 linkage. A few companies, such as Robart, Du-Bro and Nelson Hobby Supplies, produce hardware specifically for giant scale, but there are many questions on how to make everything work reliably and, above all, be strong enough for the job.

SHORT PHYSICS LESSON
Most full-scale aircraft have balanced control surfaces that, in level flight, compensate for gravity's pull on them. To achieve this balance, the surface usually has some type of weight and arm attached to it or somewhere in its linkage system.

The advantages of having a balanced surface are:
• Less force is needed to physically move the surface (easier on the servo).
• The surface resists inadvertent movement or instability, i.e., there's no chance of aerodynamic flutter. Unfortunately, at best, balancing our models' surfaces isn't easy, and it's often quite impossible, owing mainly to the large amount of throw used by most builders—including me! We rely on our servos and linkage systems to compensate for an out-of-balance situation and to be strong enough to absorb the sometimes enormous stresses to which we subject them. Keep in mind that when the airframe is stressed, the surface link is stressed even more.

BACK TO BASICS
Knowing your piloting skills and understanding the capabilities of your airframe can go a long way toward putting you in the ballpark as far as learning which equipment

The linkage setup in the fuselage: everything is neat and in order. The rudder and tailwheel servos are in the center, and the elevator servos are on the outside. Note the springs on the tailwheel-servo arms; they relieve the servo of stress and side loads.

CHAPTER 3

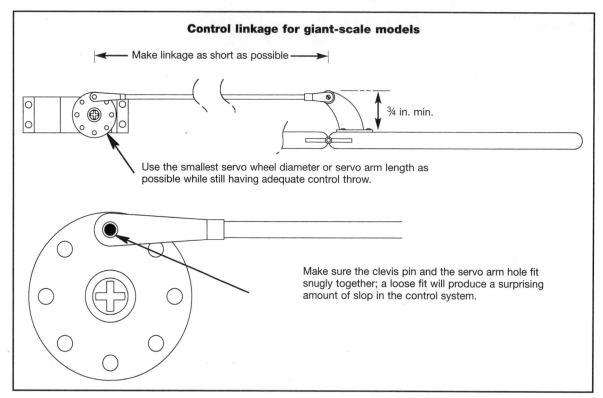

Control linkage for giant-scale models

Make linkage as short as possible

¾ in. min.

Use the smallest servo wheel diameter or servo arm length as possible while still having adequate control throw.

Make sure the clevis pin and the servo arm hole fit snugly together; a loose fit will produce a surprising amount of slop in the control system.

you need. Because of the lack of useful test data on components, most of us lean toward overkill. This is definitely adequate, but it can also be ugly and expensive. I assess the linkage needs of airframes keeping these four buzzwords in mind:

—short; —simple;
—tight; —leverage.

- **Short.** I mean as short as possible. The longer the rod, the more flexing there will be in the system and, therefore, the greater the likelihood of slop.

- **Simple.** This defines itself. Stay away from bellcranks and multiple rod links. Whenever you have to make a connection, you create slop. There has to be clearance between the drive pins in the clevis and the horn, and they can add up to a lot of play.

- **Tight.** If you took care of the first two items on my list, then you have "tight" taken care of except for the servo. Use good, tight servos that have dual ball bearings and preferably a good solid gear mesh. If the servo is sloppy out of the box, it will only get worse—usually, in a very short time.

- **Leverage.** Having proper leverage is very important—almost a subject in itself. It can also be hard to come by—especially if you build for scale competition and want to hide all of the pushrods. It's easy to get stuck within the confines of a cramped fuselage tail section or an outer wing panel where there isn't much room for a long control horn.

Short control horns provide little leverage and can lead to flutter and undue wear and tear on the system. This makes the servo work harder and puts it under load more often; this is also a big current drain on the battery. The rule for servo arms is exactly opposite that for control horns: farther out on the arm means less power. Here's a good rule to follow: couple a short servo arm with a long surface horn. This will provide maximum servo power and adequate leverage on the surface.

NUTS AND BOLTS

Most of the available "giant" hardware will do a fine job if used properly, but because I'm asked so often, I'll pass along my preferences. The pieces I mention are the ones I use in every project and have had good luck with over the years.

For throttle, I start with a good ball-bearing sport JR 531 servo and an ⅛-inch Sullivan flexible Nyrod with a Sullivan clevis at the servo end and a 2-56 ball link at the carb end. Keep in mind that for consistent response, throttle linkages should be just as tight and neat as those for the elevator.

For aileron, I start with a medium-power (90 to 100 oz.-in.) JR 4131 ball-bearing servo and a threaded 4-40 solid rod with a Sullivan clevis at the servo end and a Robart clevis and ball-bearing horn (at least ¾ inch long) at the surface end. After I've tested the setup, I solder both ends to eliminate the need for jam nuts.

CONTROL-LINKAGE SETUPS

Most IMAC and TOC airplanes are set up in this way: externally mounted servos and short, heavy pushrods. Although it isn't pretty, it's great for 100-percent power transfer, and it's easy to check.

use a reversed servo (available from JR) on one side. This allows the same side servo rotation and keeps the pushrod lengths identical. I use a JR 4721 (120 oz.-in. torque) with a straight, 4-40 rod and the longest horn the wing thickness will allow. Maximum leverage is imperative here. The flaps need to have good authority; if they don't, they can easily become asymmetric, i.e., they don't come down the same distance, and this is no fun!

I use the same type of servo for the elevator, but I use the large, 3/8-inch Sullivan Nyrod with 4-40 rods at both ends. I again use a Sullivan clevis at the servo end and a Robart clevis and ball-bearing horn on the surface (again, at least 3/4 inch long), and once again, I solder both ends.

For ease of installation and setup, I always separate the rudder and tailwheel (or nose wheel). Using one servo for both usually causes one or the other to be short on throw or reversed on both. Because the rudder is the largest control surface and the tailwheel takes a lot of abuse, I use pull/pull on both. By doing that, I ensure that they're solid and tight in both directions, and I can get maximum throw in both directions, too. The Sullivan clevis with rod and eye ties off the cable on the servo end, and I use a ball link at the surface end and drill through for the cable tie. I use the same setup for the tailwheel, with the exception that, at the servo end, I use a small, 1-inch-long, no. 2 coil spring for each cable to provide some flexibility and to eliminate cable stretching.

Strong, solid servos are needed for the flap system, and pushrod geometry must be identical on both sides. Always use one servo for each section; having several flap linkages is a geometric nightmare. A sure-fire way to simplify the geometry is to

HINTS

• When you solder a clevis that is threaded on, use extra flux to make the solder flow into the threads; don't let the solder build up on the end. Heat the joint only long enough to do the job: too much heat will weaken the steel clevis.

• Nyrods should be supported well—usually at both ends and at least once in the middle. Also, never allow any of the inner (plastic) rod to exit the outer tube; always have steel rod exiting the tube to the link.

• Before you glue the Nyrods into place, scuff the surface with 80-grit sandpaper to give the glue some bite. This is usually the only place I ever use epoxy in construction because CA doesn't stick to plastic.

• Clevis and servo-arm connections are often sloppy because of the difference between the clevis-pin's diameter and the factory-drilled hole. To eliminate this slop, carefully measure the clevis pin, then drill your own hole in the servo horn to match the pin. It's amazing how tight this connection can be and just how much slop there often is to begin with.

Being connected safely and securely isn't weird science; it's just good, commonsense mechanics.

CHAPTER 3

Servo-operated Fowler flaps
by Bob Almes

A good aerial photo of a full-size P-38 Lightning coming in for a landing. The full-size P-38 deployed 40 degrees of flap in the full-down position. In model use, Fowler flaps need only be deployed 25 degrees for the full-down position.

PHOTOS BY RICHARD HOSE

In July '92 and June '93, *Model Airplane News* published articles on my pneumatically powered three- and two-position Fowler flaps. These articles contained photos, plans (FSP07922 and FSP06932) and construction notes for building and installing these flaps in a P-38 Lightning. Since that time, I have searched for ways to simplify and improve my designs. Of particular significance is the development of a new flap hinge block that eliminated 16 items from the original parts list. When building four separate flap sections for a P-38, there are 64 fewer parts to build.

SERVOS INSTEAD OF AIR

I have never been completely satisfied with the use of air to operate flaps. Air cylinders are heavy and require a substantial amount of onboard support equipment. During one of my experiments, I was pleased that I could program auxiliary channel servos to traverse a 130-degree arc. It became apparent that a 2¼-inch long servo arm could produce 4 inches of linear travel. At the outset, the idea of such a long servo arm was a bit disconcerting. However, one of the inherent features of Fowler flaps is that they require very little power to operate. Unlike other flap designs, air loads are not directly applied to the actuating power source. In fact, air loads assist in the operation of Fowler flaps.

My friend Ed Newman is an avid scale enthusiast and a regular participant at Top Gun and the Scale Masters. Ed had acquired a set of *Model Airplane News* Fowler flap plans and contacted me about fitting the flaps into a ⅙-scale P-38. I told him about my new servo-powered version, and 40 minutes later, he was at my door. We looked over the plans and discussed the various installation options. Ed brought up the possibility of using the servo arm to provide flap rotation as well as flap extension. I was intrigued with the idea and thought it worth further investigation. Ed is a draftsman by trade and offered to draw up the plans if I worked out the details. The results are presented here.

Figure 1 represents my updated version of the original flap design. Telescoping tubes replace the original's air cylinders, a flap hinge block replaces the previous complex hinge assembly, and a servo replaces air as the power source. I used a set of these flaps in my ⅙-scale P-38 "Yippee" using JR 517 standard-size servos. The installation is lighter, less troublesome and eliminates the onboard air support equipment.

Figure 2 illustrates the concept of

Here's my P-38 model "Yippee" with the flaps fully deployed. The model lands like a butterfly with sore feet.

102 WORKSHOP SECRETS

using the servo arm to create the flap-down angle as well as flap extension. The rotary motion of the servo arm is converted to linear motion by the pushrod connected to the flap hinge block. A second pushrod is connected to the flap control horn. The flap-down angle is created by a ¼-inch difference in the linear travel of the two pushrods. This translates into a flap-down angle of 25 degrees using a ¾-inch flap control horn. The servo arm design brings into perfect harmony the three motions involved.

TUBULAR SUPPORT

I have experimented with various combinations of tubes to find the ideal material for friction-free operation. Brass and aluminum are not suitable for this application. The ideal combination proved to be Dave Brown Products fiberglass pushrods for the outer housing tubes and aligned fiber composites (AFC) 0.240-inch carbon-fiber tubes for the inner flap support tubes. If not available at your local hobby store, the AFC tubes may be obtained from CBA Models of Warren, OH.

Considering low-budget operation and the need for servo operation synchrony, I tested the JR 517 (40 oz.-in. torque) servo. During the tests, the 517 provided the desired linear travel; however, I had to sort through six additional servos before finding three that would traverse the same degree of arc as the first servo. Many low-cost servos do not maintain equal component tolerance levels, and this affects their ability to operate in complete synchrony. It appears that any standard, mini, or low-profile servo in the mid to upper price range will do the job. In some cases, wing thickness limits the size of the servo that can be used. I also tested the JR 321 (29.2 oz.-in. torque) miniservos in a ⅛-scale P-38 and encountered no problems in range of motion or power.

CONSTRUCTION

Figure 2 illustrates the general layout of the parts for a Fowler flap. No specific dimensions are given because the various wing structures require a different installation treatment. The photos of Ed Newman's ⅙-scale P-38 show how the wing structure at the inboard flap position is different from that at the outboard flap position. Having determined how and where the parts would be placed in the wing structure, Ed made a jig to hold the tubes in proper alignment while he assembled the flap support tubes and flap hinge block. The tubes and hinge block assembly then became the jig, which aligned the outer housing tubes within the wing structure while they were cemented in place. Note that the tubes supporting the inner flap

These two photos show Ed Newman's flap installation in his P-38. Note that the outboard flap installation (above) has the tubes attached to the wing ribs while the inboard flap assembly (below) has tubes supported by balsa platforms placed between the wing ribs.

assembly are mounted on platforms spaced between the ribs. On the other hand, the tubes of the outboard flap assembly are mounted on balsa rails cemented to the rib structure. Ed plans to use the JR 321 miniservo to power his flaps.

The only part of the project that requires some degree of accuracy—the servo arm—is illustrated in Figure 3. Note that 3/32-inch ply is used for the servo arm, but fiberglass, carbon fiber, or plastic may be used for this purpose. If 3/32-inch ply is not readily available, laminate pieces of 1/32-inch and 1/16-inch ply together. Cut and shape the four servo arms, then sandwich all the parts together for the hole drilling operation. Any variation from the specifications cited in the drawing will at least be reflected in all servo torque arms. The shoulder around the splined hole of the servo wheel is not the same size for all brands. It may be necessary to chamfer the ⅜-inch-diameter hole to accommodate the servo wheel being used. After the 2-56 ball link and no. 2 button-head screws have been installed, snip off and grind down the excess lengths.

Figure 4 illustrates a method of constructing the flap hinge block. Another way would be to use hard balsa or basswood shaped to accommodate the trailing edge forward sweep and holes drilled for the tubes, hinges and flap pushrod connector. This would surely require a template to ensure proper alignment of the holes. Be creative.

PROGRAMMING FLAP OPERATION

Select one of the flap assemblies as Flap 1. This flap will be used to program the servo travel endpoints. Flaps 2, 3 and 4 will later be mechanically adjusted to coincide with the operation of Flap 1. Remove the servo arm to be sure that it does not accidentally contact some part of the wing structure. A small, standard-length servo arm may be used to observe the operation.

Turn on the radio system, and enter the menu required to program the flap endpoints. Program the transmitter until the servo reaches its maximum flap-up travel. Reposition the small servo torque arm to assume the position as indicated in Figure 2. Program the servo's travel to the maximum flap-down position. Remove and reposition the servo arm as often as necessary until the flap-up and -down endpoints are the same distance from the half-flap position. The lengths of the servo arms for each scale size (⅛, ⅛.5 and ⅙ scale) are shown on the full-size plans and are designed to extract about 122 degrees of arc from the 130-degree maximum. This is to allow some room for adjustment at either end of the arc.

With the servo programmed to the maximum up endpoint, back it off about 4 degrees. Attach the modified servo arm to reflect the desired flap-up position, and connect the flap extension pushrod. Using a ruler for measurement, extend the flap to the required linear travel. The flap-up and flap-down endpoints have been established. Save these values in your computer radio and return the transmitter to the operate mode.

Adjust the pivot point of the flap hinge and the flap control horn pivot point so they are ¾ inch apart. Connect and adjust the length of the flap angle pushrod so that the flap is set in the proper flap-up position. All things being equal, the flap should reflect a 25- to 27-degree down angle when the flap is fully extended. Connect flaps 2, 3 and 4 to the radio system. Position the servo arms to reflect the same flap-up posture as Flap 1. Adjust the flap extension pushrod to reflect the flap-up posture. Adjust the flap control horn to reflect the same ¾ inch between pivot points as before.

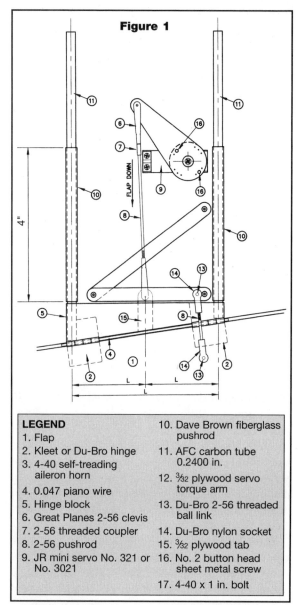

LEGEND
1. Flap
2. Kleet or Du-Bro hinge
3. 4-40 self-treading aileron horn
4. 0.047 piano wire
5. Hinge block
6. Great Planes 2-56 clevis
7. 2-56 threaded coupler
8. 2-56 pushrod
9. JR mini servo No. 321 or No. 3021
10. Dave Brown fiberglass pushrod
11. AFC carbon tube 0.2400 in.
12. ³⁄₃₂ plywood servo torque arm
13. Du-Bro 2-56 threaded ball link
14. Du-Bro nylon socket
15. ³⁄₃₂ plywood tab
16. No. 2 button head sheet metal screw
17. 4-40 x 1 in. bolt

Connect and adjust the length of the flap-down angle pushrod to reflect the proper flap-up posture. The operation of all flaps should now coincide.

RECEIVER HOOK-UP

As previously stated, Fowler flaps require very little power to operate. A double Y-connector system can be used to connect all four flaps to the receiver flap channel. This hook-up technique requires that all servos be installed so that the direction of travel is the same for all. If for any reason a flap or flaps should stall, the electrical load on a single channel may not bode well for the receiver. As a general rule, I never connect more than two servos to the same receiver channel. To keep servo leads as short as possible, I use two receivers in my P-38s. All servos on the left-hand side of the aircraft are connected to the receiver in the left boom, and all ser-

CONTROL-LINKAGE SETUPS

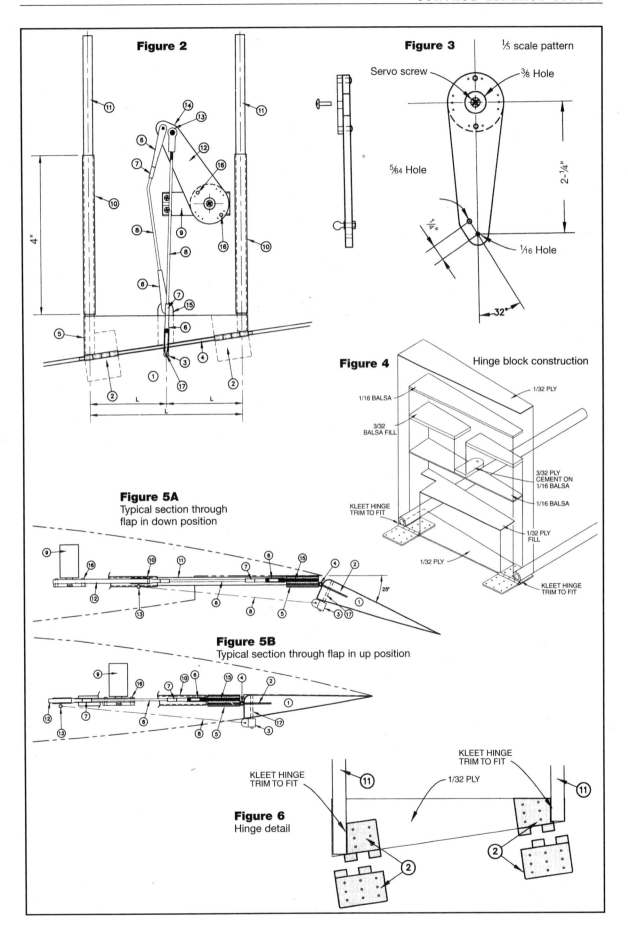

CHAPTER 3

This is Ed Newman's simple jig that he used to assemble and position the tubes before they were installed in the model's wing.

When the landing gear are extended, the landing gear bay areas, as expected, create a significant amount of drag. A 20-degree, half-flap selection caused the model to balloon. On the final approach with a full 40 degrees of flaps deployed, substantial power was required to drag the model to the runway. This was completely unacceptable and can be likened to driving with both brake and accelerator applied.

After all, we have scaled down the size of the aircraft, the flight control surfaces and the extension of the flaps. So why insist on maintaining a flap-down angle consistent with the real aircraft? Experimentation reveals that the ideal full-flap configuration should be about 25 degrees. Flight performance appears to closely approximate that of the real aircraft. Speed checks with a radar gun reveal the full-throttle speed in level flight is

vos on the right-hand side of the aircraft are connected to the receiver in the right boom. Of course both receivers are on the same frequency and the antennas remain inside the booms.

The two flaps per side can be connected to the receiver via a Y-connector or to separate channels that are mixed together. In the case where one receiver is used, it is recommended that the right-hand flaps be connected via a Y-connector to one auxiliary channel and the left-hand flaps be connected via a Y-connector to a different auxiliary channel. The two channels are then mixed together. These latter two techniques offer an added advantage in that the direction of travel for the right-hand servos can be opposite that of the left-hand servos. Appropriate direction of servo travel can be controlled by the transmitter's reversing switches. This allows more freedom of servo installation for each side of the aircraft.

Joe Grable is scratch-building a C-130 Hercules and has included the P-38 flap design to fit his model. Originally, Joe had decided to use my older, pneumatic-powered flap assemblies but decided to wait for the servo-powered version. Joe's custom-built units are completely assembled to include the servo, prior to the installation, in the wing. When completed, the total flap area for this model will be 308 square inches. That is a substantial increase in wing area for landing.

about 97mph. On final approach with flaps fully deployed, the speed is about 32mph. The 1/5-scale, 54-pound model touches down like a butterfly with sore feet.

FLIGHT OPERATIONS

I use a flap channel that is controlled by a programmable, three-position switch to provide flap-up, half-flap and full-flap deployment. I use 5-cell battery packs for the airborne equipment.

The full-size aircraft uses a flap-down angle of 40 degrees to be consistent with the flap-down angle employed on the real aircraft. However, speaking from personal experience, the model does not react to a 40-degree flap-down angle in the same manner as the real aircraft.

CONTROL-LINKAGE SETUPS

Assemble a servo power connector
by Faye Stilley

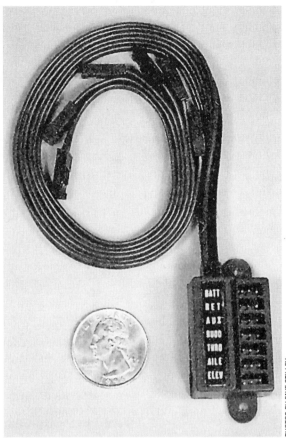

What is a servo power connector? It's a device that does good things when you plug it in between the receiver and servos. As model airplanes get bigger, servos get bigger and stronger, and more are used. Stronger, faster servos—particularly the coreless type—draw more amperage. Servo leads also get longer so servos can be close to the surfaces they control. Long servo leads create greater voltage-line loss and are more likely to transmit electronic noise back to the receiver. One stalled servo can draw 1 amp (1,000mA). Two hard-working servos can do the same. Where does that leave the receiver if the system battery is less than 1,000mAh? Sounds like an accident about to happen, doesn't it? One solution is to add a more powerful battery pack to the system. That solves part of the problem, but doesn't do anything to prevent the biggest potential hazard: electronic "noise" that travels through those long servo wires and directly into the receiver.

The best solution is to insulate the receiver from the servo power lines by using a separate battery pack for the servos. This provides more protection from electronic "noise" than any other device (short of putting an optical coupler on each servo lead). Use the 500 to 700mAh pack that came with the system to power the receiver, as most receivers draw less than 25mA. Put the big power pack on the servos. A 6-volt pack will give the servos an average 20 percent increase in speed and torque. Use whichever capacity (Ah) battery pack that you want; a 2Ah (2,000mAh) pack, or one that's even higher, will provide good insurance, although battery cells get larger and heavier as their capacity goes up. (There's always a tradeoff!)

1 The wiring diagram for the servo power connector is quite simple. Just cut the positive side of the power lead from the servo to the receiver and connect it to the servo battery. Then make a common ground between the servo battery and the rest of the system. Look at the diagram again, and multiply the number of servos times three or four for a 6- or 8-channel system. Then add a switch and a charge jack for each battery. Now picture the whole thing with wires, instead of pencil lines, connecting everything. Several questions arise. Where do you terminate the positive leads coming out of the receiver? Where do you splice in the new positive leads from each of the servos to the servo battery? Where do you splice in the new common ground wire? Don't give up! The unit can be built without a single spliced wire and with virtually no additional wire. The unit is then plugged into the wiring between the receiver and the servos.

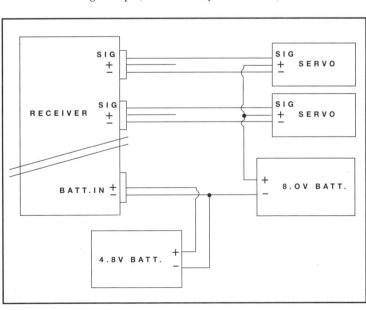

WORKSHOP SECRETS **107**

CHAPTER 3

2 I started construction with old receiver parts from a pre-1991 non-gold-stickered receiver—worthless to most people, but invaluable for this project. If you don't have an old receiver, you can probably get one from another modeler. The PC boards in these old receivers are of high quality; the case is made to fit a certain brand of servo plugs; the pin connectors are gold-plated, and most of the required soldering has already been done.

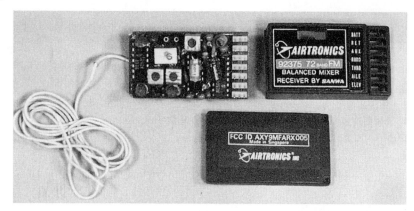

3 Here, I've "de-populated" the part of the board that I will use. After I had removed the components, I cleaned off the excess solder with a woven copper braid (wick), which is made for removing solder and is sold in all electronics stores. These braids work so well that they "suck" the solder completely out of the holes in the connector pads and leave a perfectly tinned pad ready for soldering. Look closely at the board, and you can see that all the center pins are connected to a buss. All of the pins on the right side are connected to a second buss; these are the positive and negative power connectors. This saved me from making 14 separate solder joints. Alongside each signal pin is a pad connected by a lan to the individual signal pins. That saved making another six solder joints, and the pads make it a snap to solder the new connections.

5 I glued the case parts together with thick CA. By removing the excess material from the center of the case, I retained the mounting lugs and screw posts on each end. The "new" case goes together as smoothly and tightly as the original. I carefully pried the label off the original case, cut it to size and reinstalled it on the new case.

4 The picture shows the case, board and bottom cover cut to final size. To conserve weight and size, I cut away about ⅔ of the case, leaving just enough room for the internal wiring. You can also use the entire receiver case and eliminate this step.

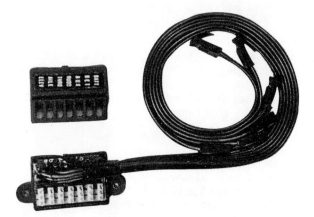

6 The leads, which are plugged into the receiver, are soldered into place. A piece of shrink-tube is used to bundle them together and provide some strain relief where they exit the case. Assembled leads with plugs and 12-inch pigtails are available from several aftermarket manufacturers. Most have gold-plated contacts, and some are available with heavier 22AWG wire for $2 to $3 each. I made the mounting tabs on the case from the scrap material I had removed earlier. They aren't absolutely necessary; you can also wrap the case in foam and anchor it in some fashion.

CONTROL-LINKAGE SETUPS

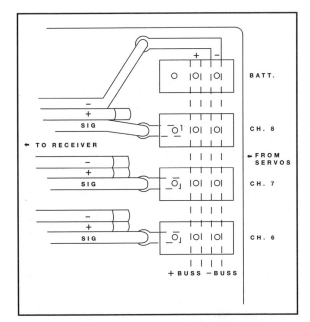

7 There's not much work involved in making the wire connections. Each lead has its signal conductor soldered to a pad alongside a signal pin. All of the leads—except one—have the positive and negative conductors capped off with shrink-tube. One lead has the negative conductor soldered to a pad that's connected to the negative buss. This provides a common ground for the whole system. (Note: some systems use the center connector for ground.) When the servo battery pack is plugged into the power connector, it is automatically connected to all the power connectors for the servos as well as the common ground. That's the whole thing—except for the switch/charge-jack installation.

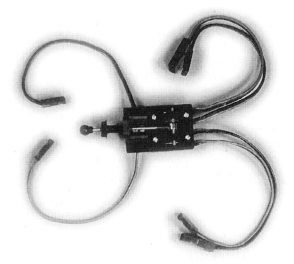

8 What could be simpler than to use a regular switch harness on each battery? A single switch to turn on/off both battery packs simultaneously, that's what. I have enough trouble with one switch, let alone trying to remember to turn on two before starting the engine and turn off two after landing the aircraft. So, I ganged the two switch harnesses by making a simple modification to a Du-Bro quick-switch mount. I cut off the inner part where the switch would normally be mounted and replaced it with a piece of plastic that was cut out so that two switches could be mounted side by side. I drilled a hole just big enough for a short no. 2 pushrod wire through each switch lever. I used a threaded piece of 2-56 pushrod wire and a wheel collar to replace the push/pull wire and plastic cap that come with the switch mount. When the threaded pushrod is tightened down through the wheel collar, the short switch-lever wire is locked tightly into place. Now there's only one "switch" outside the airplane.

Control-surface loads
by Mike Leasure

Like other advances in human history, this one started with a sharpened wooden stick. Of course, things got somewhat more complicated after that! The idea for this research came while I attempted to choose a servo size for my latest big-scale airplane, the AeroPro Laser 200. Drawing on my experiences and those of other fliers, it was obvious to me that one uses heavy and expensive servos in the 100-oz.-in. range for all controls on anything ¼ scale or larger. If it seems big, it surely must require big servos. My question is, how big is big enough, and what are the actual loads we are talking about?

I consulted a would-be aerodynamicist and modelers with extensive big-bird backgrounds, but no one could really tell me the actual loads that controls are exposed to in flight. They felt that the size of the surface and the speed of the aircraft were the deciding factors. This proved to be partially true and, to their credit, most had a good feel for the factors involved in force, but still no numbers.

THE TEST RIG

What was needed was a test rig that closely duplicated a typical model control surface and could be used to measure torque in the typical servo rating of oz.-in. The pictured rig was built and, with no small amount of embarrassment, "flown" down a local runway. The scale is a digital fish scale that's supposed to be accurate in rough water and measures in ounces. The control arm is 1 inch from the control surface center to the center of the hole in the clevis. The rig was leveled on a 1984 Ford "wind tunnel" to ensure accurate deflection angles and to eliminate turbulence from the stabilizer mounting surface. Tape was used to attach the various control surfaces tested, not only for its ease of use, but also because it provided a gap-free hinge line.

The test sequence involved running the rig down the runway at 20, 40, 60 and 80mph. This proved to be the limit of the Ford's performance (if anyone wants the data on the acceleration and

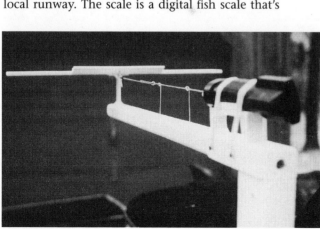

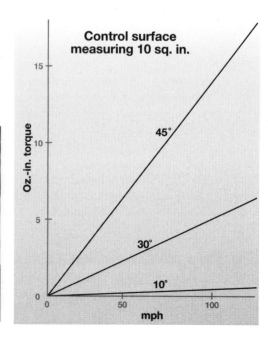

CONTROL-LINKAGE SETUPS

deceleration of a 1984 Ford pickup with a 302, I can provide that also!).

The run was made twice and averaged. The deflections used were 10 degrees, 30 degrees and 45 degrees. The surface areas tested were 10, 30 and 60 square inches, with 60 square inches duplicating one half of my Laser's elevator quite nicely.

RESULTS AND EVALUATION

The data generated was plotted on the graphs. Lines were drawn at the angle that most accurately depicted my measured results and shown at each of the three angles of control-surface deflection. Of course, 130mph is projected out from the measured data at 80mph and below. The results are depicted in a straight line, which may not be the theoretical loads, but certainly as the loads were measured—bearing in mind the crude measurement methods.

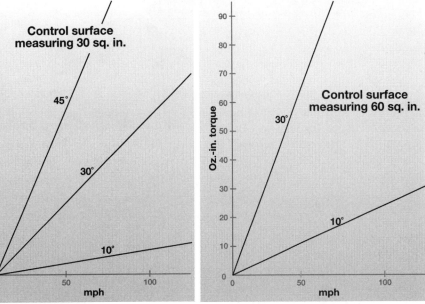

As can be seen from the graphs, speed, area and deflection play a critical role in the loads experienced. As an example, look at the result for 100mph, 10-degree deflection and 60-square-inch surface. The measured torque is only 22 oz.-in. This load is easily handled by a standard servo. The speed could theoretically increase to 160mph before the servo would be overloaded. Look, however, at the loads for the 30-degree deflection. At 40mph, I have already begun to overload a standard servo. In my sample, a surface that measures 4x15 inches is deflected 10 degrees; this actually computes to a 0.69-inch control throw. The AeroPro plans call for 0.75-inch up and down throw on the elevator, so a standard servo could be used as long as one servo was used for each side of the elevator.

The rudder is another story. Its throws more closely duplicate that of the 30-degree deflection, and a standard servo would be very quickly overcome. I will use two servos on the rudder.

Just for fun, I found the flat plate drag for 1 square foot in an old text of mine and used this to calculate the values that an aerodynamicist would project. What I found was that the same 60-square-inch surface, deflected 0.69 inch (10 degrees) and traveling 100mph, theoretically developed 29.19 ounces of drag. Stick this on a 1-inch arm, and we have 29.19 oz.-in. of force!

WEAKNESSES IN TESTING

Multiple points of improvement are possible. Turbulent airflow is a factor. Accuracy of the scale, roughness of the runway, speed of the vehicle and friction in the rig will all conspire to "adjust" results. Remember that your aircraft is also not in perfectly smooth air; it has gaps in the hinge line, and it may have less than perfect airflow patterns. This was simply my way of testing control surface loads in the real world. And now, would-be engineers and aerodynamicists, you may line up and take your best shot. I will simply say that this is what I measured on that day, at that airport, with that test rig, and these are the results. Happy landings!

CHAPTER 3

Add-on canards
by Roy L. Clough Jr.

When one of my flying buddies decided to clean house, he offered me his elderly Sig Super Kadet at a price I couldn't refuse. The Kadet was a classic—worth hanging from anybody's rafters—and it was well built (too well built for my money!). At 7½ pounds and with an ancient K&B .40 up front, its performance was most kindly described as "sedate." Naturally, I had to liven it up with more streamlined, somewhat extended wingtips and an increased aileron area. That took care of rolls, but inverted flight and outside loops the hard way (starting inverted at the bottom) escaped me. The plane was an ideal subject for the canard control surfaces I had always wanted to try. The photo shows the result of a surprisingly easy retrofit, and the drawing shows how it was done.

I feared the elevator stick would become tricky, but no problem: it was not so much more sensitive as it was more authoritative. Loops tightened up remarkably. Holding the wheels off longer, with the nose higher for a slower landing, was much easier. A real bonus: the canard's close proximity to the prop blast reduced torque effect and cut down on takeoff roll rudder diddling.

It hasn't happened too often in my design-bashing career that one of my brainstorms worked perfectly the first time (I admit to bleeding a bit for some of the innovations for which people tell me I'm notorious). But here, I've got a live one. I'm seriously considering adding canard elevators to one of my aerobatic planes. Go ahead; beat me to it!

---

5 in.
Iron-on covering

Axle is 5/32-inch wire held in a bearing made of 3/16-inch K&S brass tube that's glued into a ½-inch-square balsa crosspiece.

2¾ in.

¼-inch clearance

Put control horn on either side depending on internal servo arrangement of your plane.

Be sure to hook up canards so they work contrary to the elevators, i.e., elevator trailing edge up equals canard trailing edge down.

Canards are usually set up using a "neutral-point" formula that defines the spacing between two lifting sections. This need not apply when the canard functions only as an auxiliary elevator and not as a forward mounted wing.

11/64-inch drill-bit hole — "Eyeball" the elevator
0   1 in.   3 in.
To elevator

Use clevis rod end for trim. Attach to an inside hole of elevator arm (use about ½ the servo's elevator throw).

½-inch-square balsa bearing block

Canard control-surface incidence should start out parallel with tailplane. Tweak in elevator neutral with clevis rod end.

INSTALLATION:
Carefully drill canards with 11/64-inch drill bit and glue axle into one piece with slow-setting CA. When set up, use a couple of washers over the axle, and push it through the brass-tube bearing. Put a couple of washers on the protruding axle, squirt slow CA into the hole of remaining canard and slip the canard over the axle. Align canards before the CA dries.

4
All about engines

CHAPTER 4

Build a small-engine test stand

by Randy Randolph

Every new engine should be broken in, and most instruction manuals suggest that this be done before the engine is installed in an airplane. The best way to handle break-in is with a test stand that fits the engine. The photos show how to make a simple wooden test stand that's adequate for .03- to .20-size engines. The stand can be held in a vise or mounted on a bench and provides solid, vibration-damped mounting for glow and diesel engines.

1 Transfer these measurements to the hardwood mount and, using a coping saw or hacksaw, make a cutout that just fits the engine. In this case, the hardwood used for the mount is 1x2x6 inches because the mount is for small engines.

2 Measure across the widest part of the engine crankcase (in this case, 11/16 inch), as well as the distance from the backplate to the front of the crankcase.

3 Other than a 1/16-inch or 3/32-inch drill bit and wood screws to fit the engine mounting holes, you'll need a length of 1x4-inch hardwood, a ruler, a 1 1/4-inch hole saw, a hacksaw, or coping saw and a 12- or 16-dram pill bottle. These pill bottles are standard sizes used by all pharmacies.

4 Fit the engine into the cutout and mark the mounting holes with a pencil. Drill the mount for the proper size screws, using a 1/16-inch drill bit and no. 4 screws for small engines and a 3/32-inch drill bit and no. 6 screws for .09 to .20 engines. Treat these holes with CA to prolong the life of the mount.

5 Mark a spot 2 or 3 inches behind the mount and use a hole saw to drill a 1 1/4-inch hole all the way through the mount. This will be a good tight fit for the pill bottle, which will become the fuel tank. Attach the engine to the mount with the appropriate wood screws or machine screws.

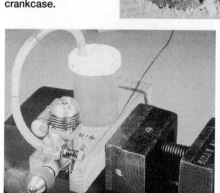

6 Drill two holes in the top of the medicine bottle's cap—one that just fits the fuel line and a small one for a breather hole. A piece of florist's wire, held in place with a rubber band behind the tank, controls the throttle. A 12-dram pill bottle will provide about 10 minutes of run time; a 16-dram pill bottle provides around 8 minutes for larger engines.

ALL ABOUT ENGINES

Build a ½A starter
by Roy L. Clough Jr.

You'll need a 6- or 7-cell Ni-Cd battery to make the starter functional.

Any used Mabuchi 550 or Goldfire electric motor, a chromed-brass, sink-drain sleeve, push-button switch and odds and ends of tubing, PVC pipe and zip cord can be used to produce this neat electric starter for .049 to .09 engines.

Remove the flux ring from the motor and the prop screw from the adapter. Clean the motor casing and prop adapter with solvent. Build up the starter cone over the prop adapter with two layers of dry surgical (or other) rubber tubing. Pry the layers of tubing apart with a toothpick and drip in a few drops of thin CA. The outer casing of the starter chuck is a short length of ¾-inch PVC pipe.

Drill a chrome-plated brass sleeve for a Radio Shack momentary-contact push-button. (Caution: thin wall tube can grab the drill and be twisted out of shape in a flash. Start with a ⅛-inch pilot hole, and bring it up to size with a succession of larger drill bits.) Slip the push-button into the hole, and

Starting your small engines couldn't be easier.

hold the nut while you turn the switch body to tighten it. Solder one side of zip cord to one motor terminal and a 2-inch wire with a prestripped and tinned end to the other. Slide wiring and motor tail first into the flanged end of the brass sleeve. Solder the motor wire to one side of the switch and the cord wire to the other. The clearance between the motor and the sleeve i.d. is taken up by rolled-up, manila file-folder stock. Cut this stock long enough to insulate the wires from the case. When the motor fits snugly, lock it in place with small drips of thin CA at several points. The end plug is a plastic bottle cap. Attach your choice of battery connector with reversed polarity. Use a 5- to 7-cell sub-C pack.

If used intelligently, this starter will not damage .049 engines. Avoid hydraulic lockup, i.e., forcing the engine to spin against heavy flooding, as this might bend a rod. Prime the engine with a drop or two of fuel, and flip the prop a couple of times by hand before you attach the glow-plug clip and apply the starter.

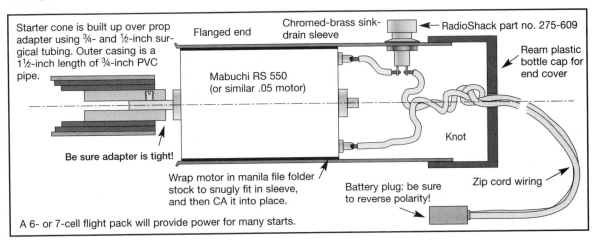

CHAPTER 4

How to make a 12- or 24-volt starter box
by Faye Stilley

Big airplanes with big engines usually need a lot of starting power. Most of us are strong enough to turn over the big engines by hand, but a leather glove won't provide protection from much more than a mosquito bite. Running on 24 volts, the Sullivan Dynatron starter produces 1.5hp, and when coupled to a Miller RC Persuader reduction drive unit, it will create enough power to turn over even the crankiest engine without endangering your hands. However, carrying two 12V batteries and a double-size starting unit is difficult without a compact storage container. This tote has a handle and a footprint smaller than this magazine page.

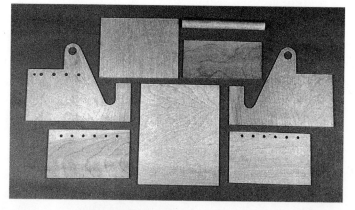

1 The tote is made with only seven pieces of ¼-inch plywood and one ¾-inch dowel. The two pieces at the top edge will enclose the battery and have a uniform row of ventilation holes that allow heat to escape. Both side pieces are drilled for the dowel handle, and one has holes for the banana plug jacks and switch. The other three pieces will make up the box floor, the battery compartment and the cradle end of the tote.

2 For simplicity, all of the plywood parts are 7¼ inches long. Set the saw, and cut blanks for all of the parts at one time. Note: if you are using imported plywood, it is probably 6mm thick instead of a full ¼ inch (6.35mm). If so, only the length of the floor and the height of the divider and battery end need to be adjusted; all of the other measurements can remain the same. Drill ¼-inch holes in the divider and battery end piece above where the tops of the batteries will be. Their placement isn't critical, but they are necessary to provide good ventilation. The holes in the side panel should fit whichever banana plug jacks and switch you choose.

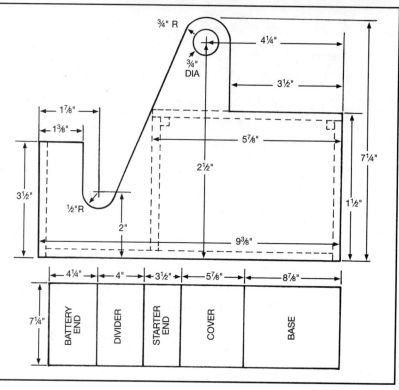

116 WORKSHOP SECRETS

ALL ABOUT ENGINES

3 Here, the tote is assembled and ready to be wired. It is important to keep everything square during assembly, or the battery cover will not fit properly. A good wood glue or epoxy will hold things together just fine. Of course, screws or nails could be used instead.

4 Mount the cover using two pieces of ¼ x ¼-inch spruce, drilled and fitted with ⅛-inch dowels. Note the holes drilled in the divider to accept the ⅛-inch dowels. Finish attaching the cover by putting a screw through the battery end of the tote and into a piece of ¼-inch-square spruce on the other side of the cover. Aside from screws, the cover can be secured in any number of ways, except with hinges: the handle will interfere with the operation of a hinged cover. The handle location shouldn't be changed because it's at the tote's center of balance.

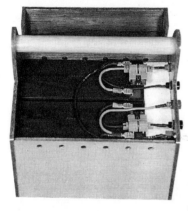

5 Here, the batteries and wiring have been installed. Wedge a piece of Styrofoam below the banana plug jacks to hold the batteries in place. Each battery is protected by a 30A automotive fuse. The "clamp-on"-type fuse holders are available at most automotive supply stores. If the fuses you buy come in a plastic bag, there is room to carry an extra supply on the top of the batteries. If they come in a metal tin, carry them somewhere else, such as the starter cradle.

6 The wiring is straightforward. You basically have two 12V power supplies that are wired so that each can be charged separately with an ordinary 12V charger (with the switch in the "off" position). Turning the switch "on" simply puts the batteries in series, providing a 24V supply. Note that the fuses protect the batteries when they're in series and parallel. Be sure to choose a switch with an amperage rating that's higher than that of your batteries.

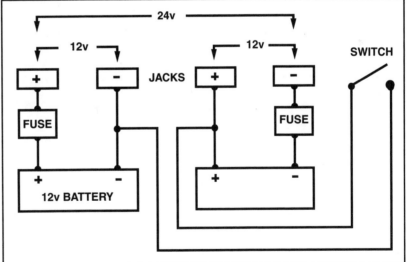

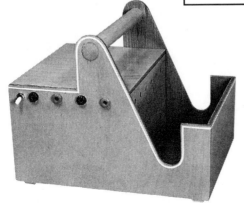

7 The tote is ready. For 12V operation, plug the starter into the first set of jacks. The second set of jacks provides a 12V backup. When you need 24 volts, plug one banana plug into the first jack, the other into the last jack, and flip the switch on. This tote is ambidextrous; the starter unit can be carried in the cradle facing either side. Here, it is ready to be used on your left side, making it convenient for "lefties." I added rubber feet to keep the tote off the ground.

WORKSHOP SECRETS 117

CHAPTER 4

Drill and tap engine mounts
by Gerry Yarrish

Aluminum engine mounts are a great invention; they're strong and light, and the new Du-Bro vibration-reducing motor mounts help to reduce aircraft harmonics and airframe-induced noise. Many of the plans in our "Plans Mart" use this type of engine mount. Here's the quick and easy way to drill and tap aluminum engine mounts.

4 Clamp both engine-mount beams in a vise with their front edges aligned, and mark the holes' centers on the second beam. A small square or straightedge is best for this.

1 First, decide which engine mount is best to use with your engine. Most engine mounts come in many sizes and lengths. Use your plan's side view to determine the distance between the firewall and the prop-drive washer. Buy the proper mount for the job.

5 Center-punch the hole locations. Take your time, and make sure you have precisely placed the punch before you strike it home.

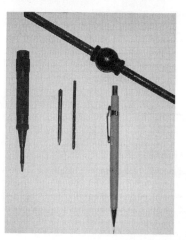

2 The tools you need: a center punch (here, I have a spring-loaded automatic punch that doesn't require a hammer); a tap and drill bit of the proper size and tap holder; a small square (not pictured); and a scribe, awl, or pencil to mark the hole locations.

3 Place the engine (here, an O.S. 1.20) on one of the engine-mount beams, and mark the engine-mount holes' position. Use a very thin pencil or scribe to accurately transfer the mounting-hole positions to the beam.

6 As you drill the holes, be sure to keep the bit square with the top surface of the beam. For an 8-32 tap, use a no. 29 (0.136-inch) bit. Drill at a low speed—about 500rpm. Back the drill out frequently to unload the bit of metal chips and to minimize heat.

ALL ABOUT ENGINES

7 Next, clamp each of the beams in a bench vise, and tap the holes. Keep the tap perpendicular to the top surface of the beam and, with firm pressure, insert the tap with a half twist. Once the tap starts to grab, keep it straight and advance it ½ turn at a time. Before you advance the tap again, back it out ¼ turn to break the metal burr that forms in the hole. If you like, you can use kerosene as a cutting fluid, but aluminum can be tapped dry while using small-size taps.

8 Clean the holes of burrs and metal chips, and mount the engine with 8-32 cap-head screws. These should fit the holes perfectly, but if they don't, you can ream the holes slightly for a proper fit.

9 With the engine-mount beams firmly attached to the engine, it is much easier to mark where the holes for the mount will be drilled in the firewall. Use vertical and horizontal center lines to obtain the thrust center line and the positions of the four mounting bolts. Drill the holes, install the blind nuts, and bolt your engine into place.

Drill bit and tap size chart

For properly formed threads in a tapped hole, you need to start with a drill bit. Here are some common bit and tap sizes. The sizes given produce a hole large enough to produce a 75-percent thread size. This is more than strong enough for model aircraft use, and it minimizes the force required to cut the threads.

Tap size	Drill bit no.	Drill decimal size
2-56	50	0.070
4-40	43	0.089
6-32	36	0.107
8-32	29	0.136
10-24	25	0.150
10-32	21	0.159
¼-20	7	0.201

■ Securing threaded joints

A threaded joint that comes unfastened because of vibration or other environmental influences can cost you your plane or helicopter. I'm sure you have seen mufflers fall off engines in flight, props fly off the drive shaft and engines come loose from their mounts. Such things happen because a threaded nut on a screw or a screw in a tapped hole tends to turn when subjected to vibration or impact. This is a common but largely unaddressed problem. If you want to secure your model's threaded joints effectively, here's what you need to know.

Mechanical locking methods

Tightening the joint puts tension on the screw to increase the torque required to move the nut. A lock washer helps keep the joint tight by increasing friction between the nut and the surface against which it bears. Lock washers are cheap but, if the nut turns enough to lessen the tension on the screw, the joint can be subject to failure. Multi-tooth washers are generally better than simple, helical spring-lock washers.

A second—excellent—solution is to use a locknut with a plastic insert that is deformed during installation. These resist being turned under vibration, even if the screw is not under tension, so they're the best choice when frequent disassembly is required (they should be replaced after a few disassemblies).

Screws with a plastic patch over part of their threads provide some of the vibration-resistance benefit of a plastic-insert locknut (at a lower price), but they must be replaced more often and aren't easy to find in retail stores.

Adhesive locking methods

Adhesives provide an alternative, and often better, solution. Here's why: "breakaway torque" is the torque required to start turning a nut; "prevailing torque" (normally less than breakaway) is the torque required to continue turning a nut. Resistance to prevailing torque is one of the big advantages that adhesives have over lock washers.

Most of us have used paint, epoxy, CA, or some other type of adhesive to secure a nut. These work fairly well under some conditions, but they don't provide a "controlled breakaway," or reliably inhibit prevailing torque, and they can shrink and be affected by fuel.

Controlled thread-locking performance for metal-to-metal and some non-metal-to-metal joints is provided by a class of adhesive that, like CA, cures anaerobically (in the absence of air). Anaerobic adhesive is frequently referred to as Loctite—the common name for the Threadlocker line of anaerobic adhesives developed and produced by the Loctite Corp. Similar materials are marketed by Loctite Corp. to the automotive industry under the name Worldtech and by Permatex Industrial, a subsidiary of Loctite Corp. Pacer Technology, the maker of the Zap line of adhesives for modelers, sells an anaerobic adhesive named Z-42 for locking metal-to-metal joints. For plastic-to-plastic joints, they market Zaplock, a cyanoacrylate (CA) product.

Pacer Z-42 Threadlocker and Loctite 242 Threadlocker. Thread-lock comes in a variety of strengths and viscosities for specific applications.

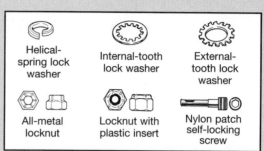

Mechanical methods for securing threaded joints

Thread-lock is most useful for locking metal, threaded joints. It's also useful for some metal-to-plastic and plastic-to-plastic applications (it dissolves some thermoplastics).

If you want to use just one grade of thread-lock for all applications, I recommend Loctite no. 242, or a functional equivalent such as Pacer's Z-42. (Use of an activator sold by the same thread-lock manufacturer helps ensure a better bond. Do not use CA accelerator.)

Thread-lock advantages
- Locking occurs in the joint and does not add to the bulk of the joint.
- If breakaway takes place, there will still be resistance to prevailing torque.
- Thread-lock cures without cracking or shrinking.
- Thread-lock will not cure outside the joint, and excess can be removed with alcohol or mineral spirits.

Thread-lock disadvantages
- Thread-lock is not ideal for joints that have to be disassembled frequently.
- The thread-lock grade discussed in this article is not suitable for use at temperatures above 300 degrees Fahrenheit.

What you need to know
- Use locknuts with plastic inserts or an anaerobic adhesive to lock threaded joints that are subjected to a high vibration or shock environment.
- For metal-to-metal joints that must be frequently disassembled, use plastic-insert locknuts.
- A 10ml bottle of thread-lock should do for most modelers and is likely to be used up within the product's shelf life—typically, one year.
- The grade you use is less important than the guarantee that it will cure; up to 24 hours is required for the joint to develop its full breakaway strength.
- An activator is required to cure a joint in the presence of some surfaces. If thread-lock is used in a critical application, especially if plated nuts/screws are involved, or if they are stainless steel or non-ferrous, use an activator.
- The safest, most conservative approach is to use thread-lock and activator on clean parts. If it's a critical application, test a joint after allowing the material to cure for at least 24 hours.

Where do you get anaerobic adhesives? You're most likely to find them at a hobby dealer, an auto-parts store, or an industrial supply house. Use them, and you'll better ensure the longevity of your favorite models. Happy flying!
—*Robert S. Hoff*

ALL ABOUT ENGINES

Make a recessed engine firewall
by Gerry Yarrish

How can you fit a 5-inch-long engine into a 3-inch cowl? Leaving 2 inches of the engine sticking out at the front looks silly, and extending the cowl often alters a model's scale outline—not good for competition. The answer is to make a recessed firewall. I've used this recessed firewall design for a while, and I think it works very well. I take a few simple measurements, and my prop and thrust washer always end up where they belong, and my models' scale outlines remain. I also make my firewalls removable to ease maintenance. I attach the fuel tank and the throttle servo to the back of the firewall to simplify installation and keep most of the weight forward (particularly important for scale biplanes and triplanes). Here's how to do it:

1 You'll need some ¼-inch-thick plywood, a scroll saw or a coping saw, screws and epoxy. Attach the engine (Saito .56 shown) to its mount and lay them on the side view of your model's plan (here, a VK Fokker triplane). To determine how deeply your firewall should be recessed, mark the position of the engine mount's rear surface on the plans.

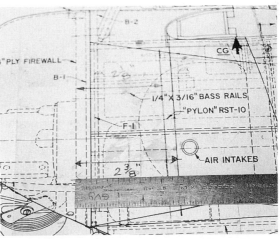

3 Now draw a second vertical line aft of the first line. This line should be the thickness of your secondary firewall aft of the first line; here, 2⅜ inches from the firewall face—for a ¼ inch-thick secondary firewall. This is also the depth of the recessed firewall box structure.

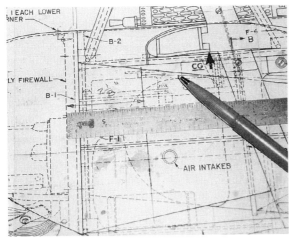

2 In this case, the distance between the main firewall's face and the rear surface of the engine mount is 2⅛ inches. On the plan, draw a vertical line there (arrowed).

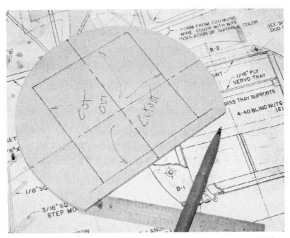

4 Establish your firewalls' centerlines, and then mark the outline of the recess. I usually make it about ¼ inch smaller all around than the fuselage's inner structure.

WORKSHOP SECRETS

CHAPTER 4

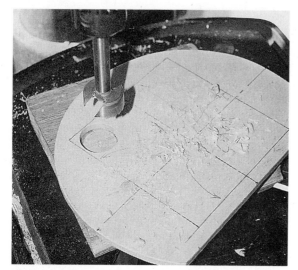

5 Use a Forstner bit or a large drill bit, and drill a hole in the inner corner of the main firewall opening. This will allow a saw blade to pass through the firewall when you cut out the rest of the opening.

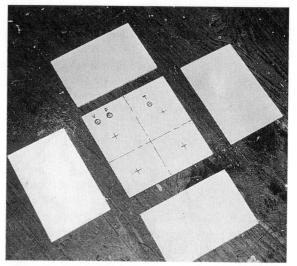

7 Cut the four walls of the recessed box out of ⅛-inch ply, and make the secondary firewall of ¼-inch ply. The walls are wide enough to fit snugly around the opening in the main firewall, and the secondary firewall (to which the engine mounts are attached) should fit snugly within the walls.

6 Use a scroll saw to finish the job of cutting out the firewall recess area. Make the corner cuts square, then lightly sand their inner edges flat and smooth.

8 The centerlines of the main and secondary firewalls should line up with each other. Now drill fuel-line and vent-line holes and holes for the throttle-linkage and the engine-mount attachment.

Epoxy the four walls to the secondary firewall to form a box, and when the epoxy has cured, fit the box into the opening in the main firewall (keep the edges of the walls flush with the main firewall face) and epoxy it into place.

9 The engine has been installed, and the main firewall has been screwed to the fuselage. Blind nuts in the fuselage match the six screws that hold the firewall in place.

ALL ABOUT ENGINES

10 To the back of the secondary firewall, epoxy a lite-ply shelf that will hold the fuel tank and the throttle servo. Note that some of the main firewall must be removed to allow clearance for the cylinder head and the intake manifold.

11 This view from above shows how nicely the tank and throttle servo fit inside the fuselage.

12 Attach the engine cowl, adjust it so that the engine fits properly through the center cutout, and then screw the aft edge of the cowl to the fuselage. Finish detailing the cowl, and you're ready to paint.

13 The finished engine cowl is dressed up with a removable dummy radial engine. The Saito muffler is barely visible, and the engine is completely hidden.

CHAPTER 4

Make a concealed muffler
by Tony Newsom

There ought to be a law against installing a big, bulbous stock muffler on a really nice-looking airplane. I've always felt that a muffler detracts from a plane's good looks, so whenever possible, I enclose it in the cowl. The plane looks better and is cleaner aerodynamically. The downside is that you have to spend some extra cash on a new muffler that may not fit your application. Here's a way to make your own mufflers and customize them to fit your model exactly.

The supermarkets are full of potential expansion chambers for mufflers. I found a tin can that was an ideal size for .40- to .60-size models such as an Extra, Corsair, P-51 Mustang and CAP 21. The juice can I used is about 4 inches tall and 2 inches in diameter. You'll need to make your own header or buy one for the engine you intend to use, and you'll need a short silicone connector. These items will be the most expensive pieces of the project, but keep in mind that barring any damage to the header, you can use it over and over on future models. In addition, you will need two tin cans, silver solder, flux, copper tube and a heat source such as a propane or butane torch. Most modelers

Left: you'll need two tin cans; the end of one will be soldered onto the pop-top end of the other, which will become the muffler. **Right:** make the inlet and exhaust holes where they best suit your application.

have everything else needed: an electric drill, round file and a tube cutter or hacksaw.

Attach the header to your engine, and mount the engine onto the plane so you can see where the expansion chamber fits best in the airframe. Use a can opener to remove the end of one of the cans. Solder this end over the pop-top end of the second can; the second can will become the muffler. Cut two pieces of ½-inch-diameter copper tube to suit your application (mine are about 1½ inches long). Use a marking pen to indicate where the inlet, outlet and pressure fittings will be positioned. Remember, you can place these tubes anywhere you like, so think about where you want the inlet and where the exhaust will exit the plane.

Use an electric drill and round file to make the holes for the copper tubes. The tubes should fit snugly into these holes, so proceed slowly and check the fit often. Use a wire brush to remove any paint around the holes down to the bare metal. Insert one of the short copper tubes into a hole and put a coat of flux all around. Use the torch to heat both the can and the tube, then feed in the solder. Solder the second tube into position, and your expansion chamber is just about finished. Make sure the solder you use is the type used in the plumbing industry—not electronic solder!

You can use the wire brush in an electric drill to remove the rest of the paint from the can and give the muffler a bright, brushed finish. You should mount the muffler rigidly to the firewall or bulkhead using metal straps cut out of thin copper sheet. Mount the header to the engine, then use the silicone coupler to join the muffler to the header. The coupler helps to isolate the muffler from vibration.

The juice-can muffler I've shown here fits inside the cowl of my .60-size Extra. The expansion chamber fits across the width of the fuselage, and the output tube exits from the bottom of the cowl. I've cut slots along the bottom of the cowl under the muffler for cooling.

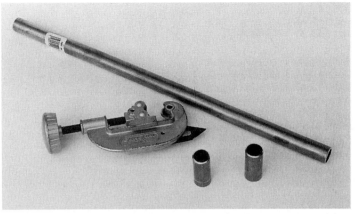
Two short lengths of copper tube will become the muffler exhaust and inlet.

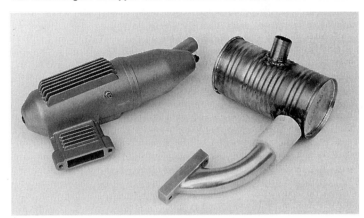
A stock muffler versus the homemade juice-can muffler.

CHAPTER 4

Part 1: Engine noise — problems and solutions
by Tore Paulsen

This article is aimed at the average modeler who operates a sport, scale or pattern airplane or maybe runs an RC car or boat. Racing airplanes—ducted fans or similar—require more work to silence, but the principles described in this article apply to most.

My "silencing" work started about 20 years ago. It was when the Merco .49 came out with a bolt-on muffler (similar to present expansion mufflers, it helped a little), but we were still losing flying fields because of noise complaints. Like everybody else, I thought that if we could develop an effective muffler, the noise problem would disappear; so I made dozens of mufflers, but my airplanes were just as noisy as ever.

I realized I needed a more scientific approach. I studied every book I could find on the subject and visited technical institutions, etc., and I started to understand the physics of acoustics. But I'll keep theory to a minimum and concentrate on the "how to" and hardware. In 1993, this magazine ran a series of articles on noise; some contained information with which I don't agree, but if you read the articles, you'll have an idea of the problems involved in assessing and quieting noise.

If, as you read this, you get the feeling that you've read it before, you might have. In '71 and '72, I was chairman of the CIAM (Committee for International Aeromodeling) noise subcommittee, and I submitted a technical paper containing the information I give in this article. It wasn't published because interest in noise reduction was too low. At the 1977 World Championship in Springfield, IL, Carl Goldberg told me that he would like to sell my muffler, but it was at least five years ahead of its time.

SOUND AND NOISE

Sound is a vibratory disturbance in a medium—in our case, air—and it spreads out (like rings in water) at, of course, the speed of sound. Noise can simply be defined as unwanted sound. To us, a sweetly running engine is music, but it might irritate the neighbors.

We're ready to establish the first rule of noise reduction: get friendly with the neighbors near your flying field! Invite them to meetings; explain the hobby and what you're doing to reduce your models' sounds; establish mutually agreed flying times; let the kids fly or taxi your models, and so forth. If you establish a friendly relationship with those who live near your flying field, they'll perceive your models as being less noisy. Caution: one loud model can destroy this relationship, and after that, even a glider may not be tolerated.

• **The loudness, or amplitude, of sound** is measured as a pressure on a specified area, such as the human ear, which can receive a wide range of sound pressures (whispers to jet-engine roars) in a non-linear way.

Sound-measuring equipment is calibrated in decibels (dB) on a logarithmic (Lg) scale; a whisper would register 0dB, and a noise that hurts would register as 120dB. It's important to understand that

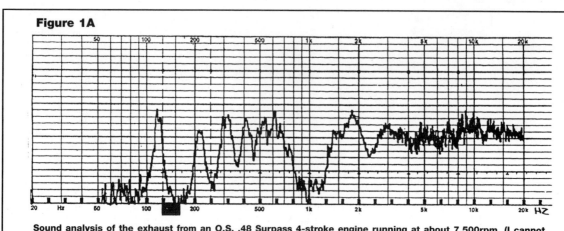

Figure 1A

Sound analysis of the exhaust from an O.S. .48 Surpass 4-stroke engine running at about 7,500rpm. (I cannot explain the low sound of 1000Hz.)

the human ear doesn't register sound linearly, and if there's more than one sound source at any time, you'll hear only the loudest sound. If you're at an airport watching jets start up or idle, and your friend tries to tell you something by yelling into your ear, you won't be able to hear him, but if the jet shuts down, his yelling will suddenly hurt your ear. This is important, because our models emit sounds from several sources, so if you manage to silence the exhaust, you'll suddenly hear the propeller—the next loudest sound—and so forth. It also means that you can't add or subtract decibel figures in the usual way; for example: two engines that would each register 90dB will produce a combined reading of 93dB on the sound meter.

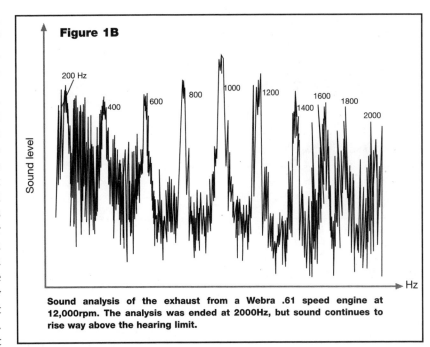

Sound analysis of the exhaust from a Webra .61 speed engine at 12,000rpm. The analysis was ended at 2000Hz, but sound continues to rise way above the hearing limit.

- **The frequency of sound waves** is recorded in hertz (Hz), or cycles per second (C/S). An engine running at 12,000rpm will have a sound frequency of 200Hz (12,000 ÷ 60 = 200Hz).

The ear does not receive all frequencies equally well. It is most sensitive to frequencies between 1,000 and 5,000Hz. To make a sound meter receive sound over this spectrum, filters are installed in it; an "A" (acoustic) is added to "dB," to denote that sound is being measured in this way, i.e., is being passed through a filter; meter readings are then expressed as "dBA."

Sound meters also have two settings: slow and fast. A slow setting damps the reading so that the meter's needle doesn't swing wildly, as it could if it were measuring the noise of a pneumatic drill. The damping effect of the slow setting averages out the meter's response to sound. For this reason, with pulsating sounds, a meter set at the slow setting may miss single peaks. Both the slow and fast settings can be used to measure a static model airplane running at steady rpm, but to measure in-flight sound, the fast setting must be used.

- **Sounds may be characterized by the shape** of their waves—sinusoidal (like a sine curve), square, or triangular. A mixture of these wave forms, such as that registered by the sound of a heavy-metal band, will make a complex shape. A soft flute tone will be sinusoidal. Our model airplanes emit a complex wave form (see Figures 1A and 1B).

- **Harmonic content.** With the exception of a pure sinusoidal wave, all sound waves contain a fundamental pitch, undertones (lower frequencies) and overtones (higher frequencies). Expressed in hertz (1 hertz—1Hz—is equal to one cycle per second) in multiples of the fundamental pitch, these overtones are called "harmonics."

Sharp-edge wave forms contain the highest number of harmonics. This is clearly shown in Figure 1B, which is an engine running at 12,000rpm or 200Hz. (Note that the sound readings given are all multiples of 200Hz.) For us, getting rid of the fundamental pitch sound frequency is very difficult; it requires the use of large mufflers and inefficient, multi-blade propellers. Fortunately, most of the sound from a model airplane comes from harmonic overtones (higher frequencies), which can be tuned out by smaller mufflers. This leaves only the basic (low) frequencies—primarily determined by engine rpm. We then hear only a quiet buzz from our models and not the screaming (high frequency) we're used to.

MEASURING SOUND

Sound (or noise) decreases with distance.
- **With every doubling of the distance** from the source of the sound, the sound is reduced by 6dB.

Measure a sound 9 feet away from its source; at 18 feet, readings will be 6dB less; at 36 feet—12dB less; at 72 feet—24dB less, and so on.

CHAPTER 4

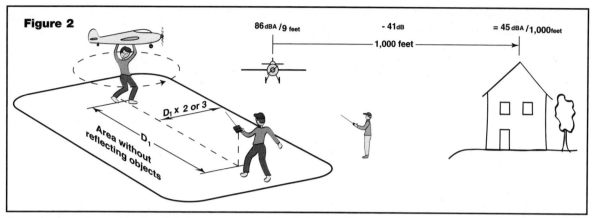

Figure 2

Expressed mathematically, this is:

$$\text{dB reduction} = 20 \times \text{Lg} \frac{\text{Distance (D2)}}{\text{Measuring distance (D1)}}$$

It's generally accepted that sound levels in housing areas should not exceed 45dBA.

• **Are you annoying the neighbors?** Look at Figure 2. Measure the distance from the air space in which you fly your models to the nearest house. Let's say that it's 1,000 feet (D2) and that you measure the sound from your models at 9 feet (D1).

$$\text{dB reduction} = 20 \times \text{Lg} \frac{1,000}{9} = 40.9\text{dB} \approx 41\text{dB}$$

Maximum sound level at the housing area should be: 45 + 41 = 86dBA

This will be the maximum sound level for your field, when measured at 9 feet. Note that the "A" refers to frequency only and not to sound levels.

I was chairman of the CIAM noise subcommittee when measuring methods were discussed. It is impossible to make accurate measurements without a laboratory, so the method now used by both the AMA and the FAI was designed to give a quick and reasonably accurate sound-level check at contests.

To check the sound level at your field with the aim that sound at the nearest house should not exceed 45dBA, this method is completely useless; there are too many possibilities for error. At our field, we tested different methods and came up with the method shown in Figure 2. One person holds the model as high as possible over his head and rotates it slowly 360 degrees with the throttle fully open. The dB meter is hand-held at shoulder height. The person measuring should wear soft clothing to avoid reflecting sound and stand at 90 degrees to the measuring line—again, to avoid reflecting sound from the chest. The test range should be permanently marked and in a quiet corner of the field.

This method has proven to be accurate when checked at the proper distance. When testing starts, the wind should be almost calm—reading no more than a maximum of 20dB below the noise limit. Measure at least 15 feet away from the model to avoid small distance errors giving inaccurate readings (9 feet is too close). On the test range, there should not be any sound-reflecting objects within 30 to 45 feet of the model.

• **What's making the noise?** Having dealt with the basic physical properties of sound/noise, I was ready to start on the hardware. But before starting to cut metal, I had to know from which parts of the model the sound was coming.

Figure 3 shows what I discovered. The values are only valid for the conditions shown. Other engines, propellers and models will give different results, but they should be in the same area. I spent a whole summer making test stands and other equipment. I will not go into details about

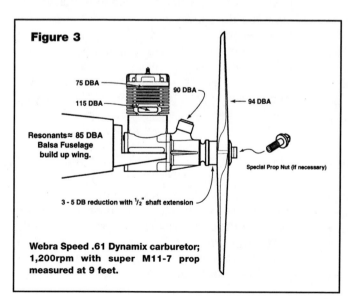

Figure 3

Webra Speed .61 Dynamix carburetor; 1,200rpm with super M11-7 prop measured at 9 feet.

the testing procedures; you'll just have to believe me, but to give you an idea of what it takes to measure exhaust sound: all other sound sources must be eliminated; the propeller must be replaced with a braked flywheel; the carburetor must be connected by means of a heavy rubber hose to a large muffler; and the engine must be mounted on a very heavy test stand.

You can now understand why muffler noise test results obtained with the engine pulling a propeller and without any reference to rpm or distance are without any value.

We will now examine each sound source and build the hardware to deal with it. It must be remembered that the object is to reduce all sounds with frequencies between 1,000Hz and 5,000Hz. These harmonic overtones give our models their screaming sound.

EXHAUST SOUND

This sound is caused by the sudden release of high-pressure gas. Four-stroke engines are a little more quiet, because the gas, when released, is of a lower pressure. But if you increase the power of a 4-stroke engine, it will quickly be almost as noisy as a 2-stroker.. The burning rate of gas in the cylinder is almost constant, so if you have an engine with a short stroke, it will be noisier than a long-stroke engine because the gas is released at a higher pressure. For this reason, an unmuffled Cox 049 has a sharp exhaust sound. In eight years, the only complaint we heard at our field was over an unmuffled Cox 049. Diesels are quieter because most of the fuel has been burned by the time the exhaust port opens; all my engines are diesels. It is no problem making a muffler that reduces sound by making it so restrictive to gas flow that the engine loses power and thereby produces less noise. Most commercial mufflers are designed this way.

Generally speaking, there are only two types of effective muffler:
• **Dissipative muffler.** This consists of a straight-through pipe that has a number of holes drilled in it and is surrounded with fiberglass, steel wool, or another absorptive material—all contained in a chamber through which the pipe runs. The sound energy is absorbed mechani-

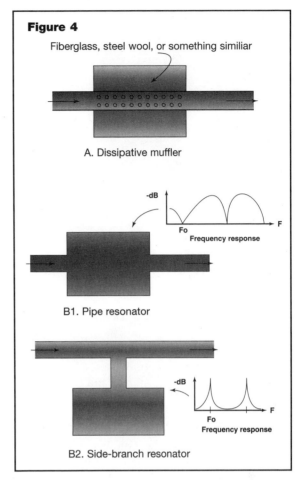

Figure 4

A. Dissipative muffler

B1. Pipe resonator

B2. Side-branch resonator

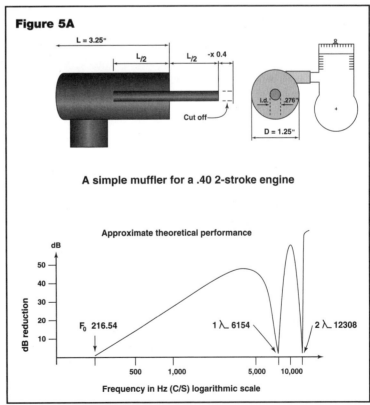

Figure 5A

A simple muffler for a .40 2-stroke engine

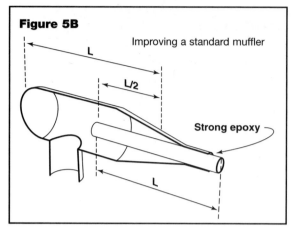

Figure 5B — Improving a standard muffler

cally by the absorptive material. This type of muffler isn't suitable for model engines because the holes quickly get clogged by the high oil content in the exhaust.

• **Reactive muffler.** This consists of a number of chambers and pipes that present a mismatch in terms of size and length in relation to the frequencies involved, and the sound energy is reflected back and forth within the muffler and is consumed there. There are two types: the pipe resonator and side-branch resonator. Auto mufflers are usually a complex combination of these mufflers. Figure 4 shows these mufflers in their simplest form.

During my many experiments, I found the pipe resonator to be the most suitable—both for performance and simplicity—so I will discuss this muffler in detail. It has one resonant frequency (Fo) and damps sound by 12dB each time the frequency is doubled (1 octave) above Fo. All chambers and pipes will have standing waves when the physical length of the pipe or chamber is an integral number of half or whole wavelengths of a frequency.

When a standing wave occurs, the sound reduction is greatly reduced. You can deduce from this that the chamber and pipe should be of equal lengths, so that we have only one set of bandpasses, and not one for the pipe and one for the chamber. To be exact, the acoustical length of the inner pipe(s) should be 0.4 x diameter, so any excess length should be cut off at the far end (see Figure 5A). If we push the pipe exactly halfway into the chamber, the half-wavelength bandpasses are canceled, and we are left with only whole-wavelength bandpasses; this greatly improves the muffler's performance. It is like a musical instrument in reverse, so the muffler must therefore be built accurately.

This may sound confusing, but if we design a pipe resonator muffler for a .40 2-stroke engine, it should be more understandable. First, the chamber volume should be at least 10 times the cylinder volume. It is best to make the muffler short and fat, so the first bandpass will be high in frequency; but the muffler should also be streamlined, so I selected a diameter of 1.25 inches.

Formula 1.

$(1.25 \div 2)^2 \times \pi \times 3.25 = 3.98 \text{ci}$

So the length of the pipe and chamber should be 3.25 inches.

The pipe's inside diameter determines the back-pressure, and a formula has been developed, by trial and error, for this diameter (except when otherwise stated, all dimensions are internal dimensions).

Formula 2.

Pipe diameter in inches =

$$\sqrt{\frac{\text{cylinder volume} \times \text{rpm}}{63{,}012}}$$

For our .40 engine, the pipe diameter comes out at 0.276 inch. We can now calculate the frequency performance of this muffler. The Fo where there is no dB reduction is equal to:

Formula 3.

$$\frac{a}{2 \times \pi} \times \sqrt{\frac{A}{V \times L}}$$

a = speed of sound in hot exhaust gas (about 20,000 inches per second);
A = cross-sectional areas of pipe:
  $(0.276 \div 2)^2 \times p = 0.0598$ sq. in.;
V = volume of chamber =
  cylinder volume x 10 = 4ci;
L = 3.25 in.;

$$Fo = \frac{20{,}000}{6.28} \times \sqrt{\frac{0.0598}{3.98 \times 3.25}} = 216.54 \text{Hz}.$$

At what frequency will we have the first standing wave with the reduction of performance? The length of the pipe and chamber is 3.25 inches. At what frequency is 3.25 inches equal to one wavelength?

Formula 4.

$$F = \frac{1 \times a}{L} = \frac{1 \times 20{,}000}{3.25} = 6154 \text{Hz}$$

The next bandpass (two wavelengths) will occur at 12,308Hz, and so forth. It should be a good muffler because we have covered the frequencies from 1,000Hz to 5,000Hz, which are the most irritating to the human ear.

ALL ABOUT ENGINES

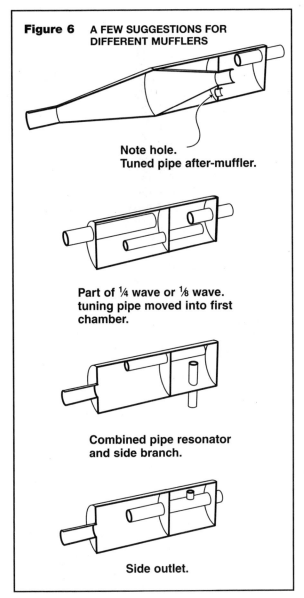

Figure 6 A FEW SUGGESTIONS FOR DIFFERENT MUFFLERS

Note hole.
Tuned pipe after-muffler.

Part of ¼ wave or ⅛ wave. tuning pipe moved into first chamber.

Combined pipe resonator and side branch.

Side outlet.

The exhaust inlet can be anywhere on the muffler, but its long internal pipe(s) can rob the engine of some power. Later in this article, I'll show you how you can connect it with a mini-pipe and increase rpm by 1,000.

The muffler and the performance chart are shown in Figure 5A. A standard expansion muffler has poor sound-reduction performance, because, although it has an expansion chamber, there is no internal pipe, or just a very short one. By just adding a pipe, you can greatly improve its performance. Figure out the diameter you need (Formula 2), and drill out the outlet to fit the pipe. Then push the pipe halfway into the muffler, and glue it with a strong epoxy. If there is not enough material to drill out, you can try a pipe with a smaller diameter (Figure 5B). For small engines, this may be the only step required to bring down the sound to an acceptable level—little cost or work. I'll explain how to make a muffler later.

VITAL PIPE RESONATOR

The reason for going into such detail about the pipe resonator (Helmholz chamber) is that this is the building block for more complex mufflers, and it will also enable you to experiment with mufflers, or custom-build one to suit special needs.

There are numerous ways to use the pipe resonator or the side branch, and some examples are shown in Figure 6. For more details, I will refer to Beranek, "Noise and Vibration Control," (McGraw-Hill). You will see that, by varying the chamber and pipe, you can design mufflers to suit special frequency needs.

If you want more sound reduction than the single pipe resonator can give, you need a second chamber and pipe. You'll probably need them for .40 2-strokers and upwards. By making the second chamber and pipe L x 0.666 inch shorter than the first but maintaining its diameter, you'll achieve a staggered effect, so that when the first chamber has a bandpass, the other chamber will have a peak, giving high reduction of sound over the entire frequency range. I have measured a reduction of over 30dB (compared with open-exhaust readings).

I sent a .60-size muffler to *Model Airplane News* to be tested by David Gierke, but he was unable to achieve exactly the same result. I do not have all the details and data of his testing, but it is very easy to get inaccurate results when measuring sound. As mentioned earlier, I spent a whole summer on obtaining consistent results. To test accurately, it's best to mount such a muffler, do the other steps required, and compare your model with others in flight.

Well, back to the muffler. I call this two-chamber muffler the "TP" muffler. Its general layout and typical frequency response are shown in Figure 7. Figure 8 shows the required dimensions—in millimeters—of mufflers in three sizes (in millimeters, to avoid my making mistakes when converting to inches; to convert to decimal inches, divide by 25.4). (Diameters aren't critical; only lengths are.) In Norway, thin-wall brass tube is available in all sizes; in the U.S., you have to look for thin-wall, 0.02-inch, brass or soft-steel tube.

Without going into tuned-pipe theory, I will just give you the length-versus-rpm formula for a normally timed engine. (These lengths are approximate.)

- A tuned-pipe length of 10.7 inches will allow 10,000rpm.
- Minus 1 inch = 1,000rpm increase.
- Plus 1 inch = 1,000rpm decrease.

WORKSHOP SECRETS **131**

CHAPTER 4

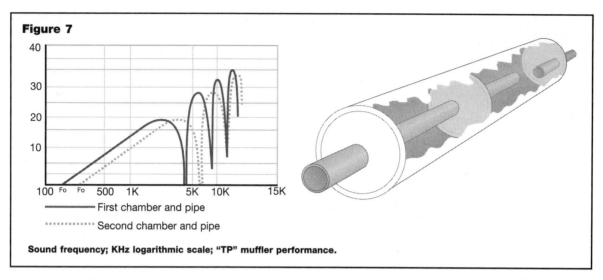

Figure 7
First chamber and pipe
Second chamber and pipe
Sound frequency; KHz logarithmic scale; "TP" muffler performance.

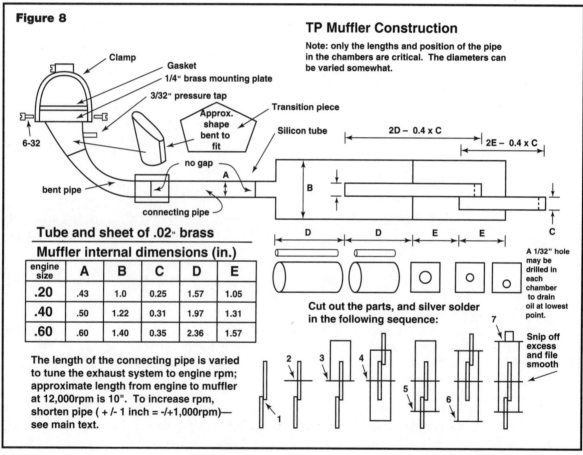

Figure 8

TP Muffler Construction

Note: only the lengths and position of the pipe in the chambers are critical. The diameters can be varied somewhat.

Tube and sheet of .02" brass

Muffler internal dimensions (in.)

engine size	A	B	C	D	E
.20	.43	1.0	0.25	1.57	1.05
.40	.50	1.22	0.31	1.97	1.31
.60	.60	1.40	0.35	2.36	1.57

The length of the connecting pipe is varied to tune the exhaust system to engine rpm; approximate length from engine to muffler at 12,000rpm is 10". To increase rpm, shorten pipe (+/- 1 inch = -/+1,000rpm)— see main text.

Cut out the parts, and silver solder in the following sequence:

A 1/32" hole may be drilled in each chamber to drain oil at lowest point.

Snip off excess and file smooth

You also have to test-fly your airplane. If the pipe is too short, you will have high static rpm, but the engine will run hot and burn plugs in flight. If it is too long, the engine will just stutter and never run right. If it is of the correct length, you will feel the engine start to pull hard when your airplane is pulling up in a loop, and it will run a little unevenly as it picks up speed coming down. Again, flying straight and level, the engine should run smoothly.

The complete system then consists of a header, a connecting pipe and a muffler, joined by silicone tubing (see Figure 9B). Make the connecting pipe longer than necessary, and use a hacksaw to cut off ¼ inch for each flight, until the engine runs as described. The tuning is fairly broad. There must be no gaps between the sections of the pipe, or the sound will escape through them and through any leaks you have in the system. Even the silicone tubing going to the pressure tap is suspect. The pressure tap can be installed in the muffler or the

ALL ABOUT ENGINES

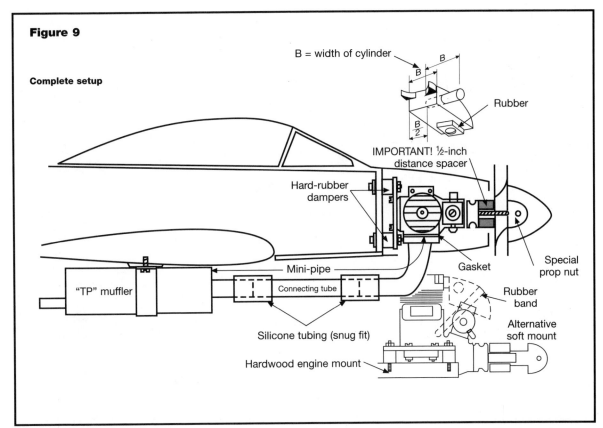

Figure 9

Complete setup

header. If this connector (mini-pipe), which is tuned to ¼ wavelength, is too long for your installation, use a mini-pipe that's half as long, tuned to ⅛ wavelength; it will also work.

You can also put the pipe inside the first chamber, but you must make a place for it by increasing the chamber's diameter. The cross-sectional area of this header and pipe should be about the same as the area of the exhaust port in the cylinder, or slightly larger than it. Avoid having sharp bends in the pipe and flat sides on the muffler.

Some sound will escape through the system's walls, and the only solution to this is more weight, so soft steel is better than brass. You can also wrap the muffler with full-size-muffler repair tape, but it isn't really necessary, because this muffler reduces sound so much that the propeller and other sources are more significant and have to be dealt with first.

In Part 2, I'll discuss how to make a "TP" muffler; the rules to follow to make it work; controlling prop and carburetor sound; and various other noise damping devices.

CHAPTER 4

Part 2: More engine-quieting solutions by Tore Paulsen

In Part 1, I discussed engine noise in general, and I gave some formulas and muffler theory. I will now discuss the sources of noise, how to deal with it and how to make the hardware.

HOW TO MAKE A TP MUFFLER
Use brass or soft-steel tube of a size close to the figures shown in the table in Figure 8 in Part 1. Cut the tube with a band saw or a Dremel cutting disk. The header is made of ¼-inch plate, shaped to fit the exhaust port. From a plumbing-supply shop, buy a 45-degree, thin-wall copper elbow that fits the connecting pipe, or make it according to Figure 9. Flatten it slightly where it meets the ¼-inch plate, and braze it on; then braze on a short pipe to the 45-degree bend.

An easy way to braze together the muffler pieces is shown at the bottom of Figure 9. For brazing, I use silver solder (which consists of 41 percent silver, plus cadmium, copper and zinc) and powder flux. First, the parts to be brazed have to be clean. Using a propane torch, heat the brazing rod and dip it into the flux; some of the flux will stick to the rod. Heat the part and touch it with the brazing rod. If the temperature is right, the rod will melt and flow around the joint. Wash off excess flux with hot water. When the parts have been joined, spray the assembly with a black, heat-resistant paint.

After you've flown your model, there will always be a little oil in the muffler, so you should always store your models nose-up to avoid getting too much oil in the engine and making it hard to start. Because of the turbulence in the muffler, almost all of the oil comes out with the exhaust when the engine is running.

The muffler for a 4-stroke engine need be only a single-chamber pipe resonator. The small-diameter pipe and the pipe that goes into the chamber should be of the same diameter as the exhaust pipe supplied with the engine. If you think that making a muffler is too difficult or too much work, they are available in five sizes, in

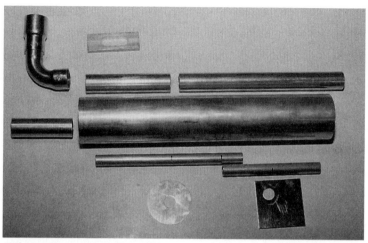
All you need to make a TP muffler.

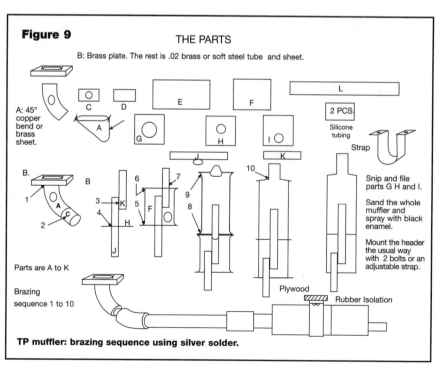
TP muffler: brazing sequence using silver solder.

134 WORKSHOP SECRETS

ALL ABOUT ENGINES

limited numbers, here in Norway. The complete system consists of gasket material, header, connecting pipe, muffler and silicone tubing.

• **Muffler rules.** To make this muffler work, you must follow a number of rules:

1. The first chamber's volume must be at least 10 times larger than the cylinder volume at 12,000rpm or lower; higher rpm require more volume.
2. The length of the second chamber must be equal to the length of the first multiplied by 0.666.
3. The internal pipes should be of the same length as the chamber, minus the mini-(header) pipe's diameter, multiplied by 0.4.
4. The internal pipes must be pushed exactly halfway into the chamber. This is important.
5. The different sections of the mini-pipe should fit tightly against one another.
6. The system must not have any leaks. To check for leaks, close the exhaust port, plug the pressure tap and blow into the outlet.
7. The connecting pipe should be of the correct length to tune the engine.

A completed muffler—painted and ready to install.

OTHER SOUND SOURCES

• **Sound from the carburetor.** This is caused by the sudden opening and closing of the intake port. The column of air going into the carburetor is frequently interrupted, and this creates a sound like an air siren. The sound's relationship to rpm does not appear to be linear.

I solved this problem by installing a simple pipe resonator (see Figure 10). The high dBA reading I got can be explained by the Webra Dynamix carburetor's plain venturi. A spraybar or a filter will lessen the sound. The best way to quiet the carb is to use a rear-intake engine; then, the forward part of the fuselage will be used as a muffler. Anyway, the trend is to use lower rpm, so that the carb makes less noise and muffling isn't necessary. You can get an idea of carburetor sound by removing the top cover of your auto's carburetor and driving around the block; your car will sound like an Indy 500 car.

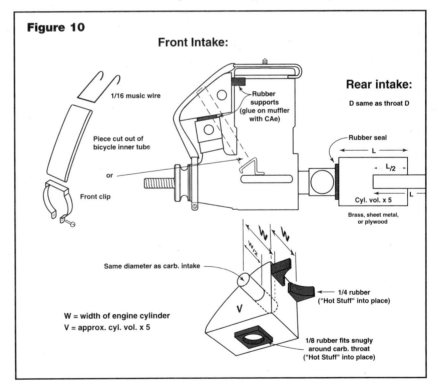

• **Sound caused by engine/airframe vibration.** I once read that if you looked at your airplane's total area as a loudspeaker and your engine as the moving coil, a .60 engine at full rpm would emit sound at around 90 watts. I decided to investigate.

I mounted the engine with a number of strong rubber bands so that it could move freely in all

WORKSHOP SECRETS 135

CHAPTER 4

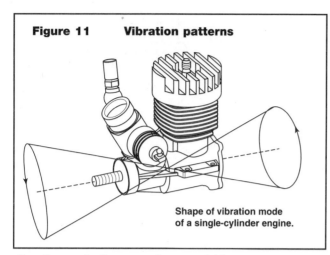

Figure 11 Vibration patterns

Shape of vibration mode of a single-cylinder engine.

We have tried to soft-mount 4-strokers, but they shook everything apart, so more research should be done in this area. Use scrap foam and balsa in your airplane to help damp airframe noise. Also look out for rattling from bellcranks, wheel bearings and other parts. Support pushrods in the middle with foam bearings. I think fiberglass fuselages will be OK if the engine is soft-mounted and large, undamped areas are supported with a bulkhead or glued to foam.

• **Sound from inside the engine.** This is already so low that little can be gained by cowling the engine, or taking any other action.

directions. I then used a variable-frequency stroboscope to look at the engine while it was running. The vibration pattern looked like Figure 11. I tried different mountings with rubber instrument dampers, and I found that a radial mounting transmitted less vibration to the fuselage than an axial mounting because it permitted the rear part of the engine to swing more freely. Full-size engines follow the same principle and use a dynafocal mount. The engine didn't vibrate much. At 10,000 to 12,000rpm, it was perfectly steady, but some vibration occurred at idle.

The soft mounts—both radial and axial—are shown in Figures 12 and 12A. The hardwood mounts can be drilled and threaded for the rubber dampers. Strengthen the threads with Zap (from Pacer Technology), and you can then screw the dampers directly into the hardwood. The nuts on the engine side need not be tightened too hard. I have flown with soft mounts since 1977, and I've never had a nut come off. Do not use Loctite; you will just twist the damper off when you try to loosen the nut.

It is important that the mounts be soft enough to allow the shaft, held by two fingers, to be moved easily about ¼ inch in all directions—from side to side and up and down. Sullivan mounts work well. I use 10mm-diameter, 10mm-long rubber instrument dampers with a threaded stud that has 4mm of threads showing at each end. The hardness should be between 40 and 60 shore.

Install the mounts symmetrically on the engine centerline according to the vibration pattern shown in Figure 11. The distance between the dampers is not critical; Figure 12 shows the approximate distance. The softness test described above will show whether you have the right dimensions.

Ready to fly: a radial- and shock-mounted engine with ½-inch spacer between the prop and the thrust washer; muffled carburetor; multiblade prop. The tuned pipe is on the rear.

• **Propeller sound.** I've saved this till last, because it's the most difficult problem to solve.

My first action was to find out how much sound came from the propeller; I ran a number of different propellers with an electric motor. The problem was that, at the high rpm needed, the electric motor was itself making too much noise, and this had to be deducted from the propeller sound, so obtaining accurate readings was difficult.

Luckily, my old friend, Jan David-Andersen, who used to make excellent diesel engines, had made an electric machine that measures both thrust and horsepower at different rpm. The whole thing ran very quietly, so I used it to check my propeller sound readings.

The propeller sound graphs are shown in Figure 13. I tried in-flight sound measurement to compare it with the static readings, but it was too difficult to position the airplane in the exact position I needed for accurate readings. Subjectively, I can't hear

ALL ABOUT ENGINES

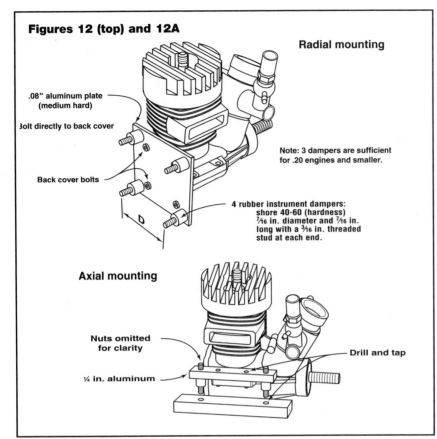

Figures 12 (top) and 12A

much difference between static and in-flight sound. The sound is caused by the pressure difference across the propeller blade. As the propeller rotates, a pulse will be heard whenever a blade passes your line of hearing; these pulses will produce a tone. Sound is also produced by objects that are so close to the propeller that they disturb the pressure pattern; this is called the "near-field" effect. On all front-intake engines, the carburetor is always too close and causes this effect. If you install a ½-inch-thick spacer between the prop driver and the propeller, the sound level will drop by 3 to 5dBA. I have included Figure 15 to help you select a prop by comparing rpm and hp. In-flight rpm will increase by about 10 percent.

Ways to reduce propeller sound:
1. Reduce tip speed.
2. Avoid having objects, e.g., a cowl or a carburetor, close to the propeller; use a spacer (see Figure 14).
3. Use thicker blades (less important).
4. Use wider blades.
5. Use more blades.

I spent a summer testing the dynamic performance of propellers, but because of illness, my tests were limited. I will discuss the results generally, but in this area, there is room for much research. *Model Airplane News* has recently included articles about this problem, but I have chosen to look at it from a different angle; if anyone agrees with me, it would be nice if they would research them from this angle and publish the results as an extension to this article. I think it will require some computer power, because the dynamics involved will be complex. Let me discuss the factors mentioned in the list above.

1. Tip speed can be reduced by reducing prop diameter and increasing pitch; however, I feel that the performance of a pattern or sport airplane will

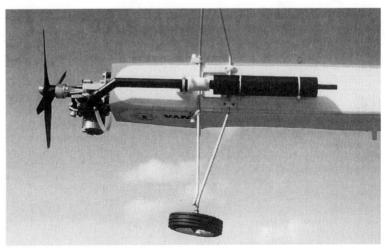

A bottom view of a complete installation minus the carburetor muffler.

be judged by its vertical behavior, and an airplane with a small-diameter high-pitch prop will quickly lose speed and maneuverability when going vertical, in spite of its high initial speed.

Tip speed can also be reduced by increasing the prop's diameter and reducing rpm. Reduced rpm will reduce horsepower, but at one point in the airplane's speed range, it will gain thrust more than it loses horsepower, so speed and maneuverability will be

WORKSHOP SECRETS **137**

maintained when flying vertical. If you also increase pitch, as is the current trend in pattern, propeller efficiency is quickly lost when slowing down. To find the best combination of propeller, airplane and engine, both to minimize sound and maximize performance, there's still a lot of work to do.

2. The near-field effect can be avoided by positioning the propeller farther forward using the spacer as shown in Figure 14 or by using an engine with rear intake.

3 and 4. There are definitive differences among the sound levels from different propellers. I can't give you a list of brands, but a wide blade with a good airfoil section is best. APC and Robbe Dynamics are good.

5. As the number of prop blades is increased, the number of harmonic overtones is reduced, so there's less sound. Prop diameter must be reduced to maintain a reasonable rpm; but as diameter is reduced, efficiency is reduced, so a 4-blade propeller, which I tried, was not the solution, even if the sound was dramatically lower. Maybe it can be used on engines with high torque? Reference 2 (see end of article) describes prop problems in depth.

THE FUTURE

One thing is certain: either we reduce noise, or we do not fly. This article has dealt with how to reduce noise with present equipment, but I believe that we have to make radical changes to the airplane, engine and propeller combination.

The rule makers (the FAI and AMA) must avoid making rules that lead to the production of noisy airplanes, and/or they must limit noise emission gradually over a number of years, until a satisfactory low sound level—and a less irritating "sound picture"—has been reached. On behalf of the CIAM noise committee, I made a proposal to this effect at the CIAM plenary meeting in 1982, but it was not accepted.

Regarding engines, there has been little development of the 2-strokers, except for increasing rpm to increase power. At the beginning, 4-strokers were quiet, but as their power was increased, they became almost as noisy as the 2-strokes.

The same will happen to electric motors as power is increased. The motors themselves do not make noise, but the propellers do. What we need is an internal-combustion engine with good power at low rpm. A refinement of the diesel engine is one solution. The PAW .49 RC engine will pull a

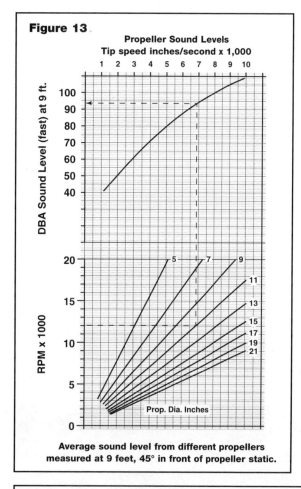

Figure 13. Average sound level from different propellers measured at 9 feet, 45° in front of propeller static.

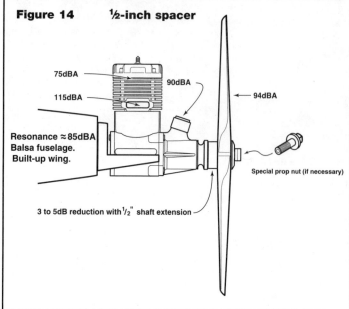

Figure 14. ½-inch spacer

ALL ABOUT ENGINES

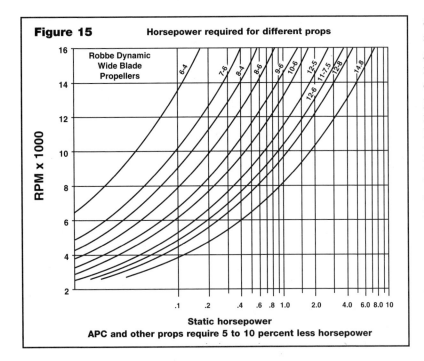

Figure 15. Horsepower required for different props. APC and other props require 5 to 10 percent less horsepower.

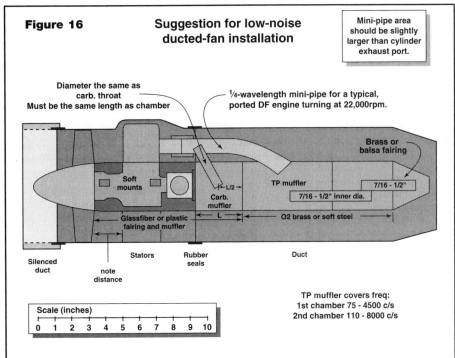

Figure 16. Suggestion for low-noise ducted-fan installation.

14x8 at 6,750rpm, which will give prop noise of about 80dBA at 9 feet (see Figure 13) and 7.5 pounds of static thrust (see Figure 15). This is comparable to a high-revving .40, but with much lower noise emission.

I have always wondered why nobody, to my knowledge, has experimented with the uniflow scavenging of a 2-stroker. It consists of a number of bypass ports at BDC, and a mechanically operated exhaust valve in the head. The valve is operated directly from the crankshaft without gearing. It will give very high torque at low rpm, allowing it to pull large-diameter, multi-blade propellers. The propeller also needs some work, but I do not know enough about this subject to offer any solutions; I know, however, that the scimitar propellers of full-size non-ducted fans are that shape to reduce noise.

SILENCING DUCTED-FAN MODELS

I had not intended to write too much about ducted fans (DFs) or racing engines, but because of the high interest in them, I will share some ideas on this subject.

I haven't done any practical experiments with DFs, so this is meant to be a basis for further work, both by the modelers and the manufacturers of these units.

First, the DF engine, which is usually of .91 size. I've stated that the first expansion chamber in a silencing system should have at least 10 times the cylinder volume at 12,000rpm to keep the backpressure and gas flow through the system within certain values. However, a DF engine runs at twice the rpm; consequently, the chamber should have 20 times the cylinder volume.

I made a DF muffler for Col. R. Thacker, but I forgot to take this into consideration and, of course, the muffler did not work (sorry, Robert!). Added to the problem of increased gas flow is that the gas is also released at higher pressure owing to the high exhaust port. The exhaust system must therefore be quite large and is difficult to fit inside the duct. I suggest a solution in Figure 16. In theory, a multi-blade DF should be much quieter than a two-blade,

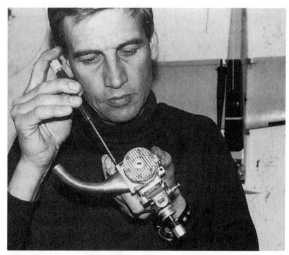

The author works on the header assembly for a tuned pipe.

5-inch-diameter propeller. At 22,000rpm, the propeller will emit 85dBA at 9 feet, but because the fan is multibladed, most of the harmonic overtones are canceled and the sound is much less—as long as there's nothing in close proximity to the fan's rotating pressure pattern. But that's just what we have! Several stators, or flow straighteners, are positioned just behind the fan, and they act like an old-fashioned hand-cranked air siren as we know them from movies of WW II. Don't take my word for it; check Reference 2 (see below), which also states that, in addition, a shroud will halve the sound level and increase thrust.

The air-intake silencing—just like the full-size fans have—is mainly to reduce the interference sound from the stators. I estimate that the stators should be no closer to the fan than 1.5 to 2 inches. If this area is smooth and free, very little thrust should be lost. In Figure 16, I have made a sketch of how a ducted fan powered by a .72 engine turning 22,000rpm should look. The fan itself is now pretty well optimized, so I will leave it as it is.

I hope this will get somebody started on making a ducted fan from which only the rush of air is heard!

REFERENCES

1. Beranek. *"Noise and Vibration Control."* McGraw-Hill.
2. A. Reiger and H. Hubbard. *"Status of Research on Propeller Noise and its Reduction."* Journal of the Acoustical Society of America, May 1953 (courtesy of R. Weber—AMA).
3. P. Fücker. *"Lärm-Lärm—Reflexions Schall Dampung Mittel Reihen-Resonator."*
4. Sweitzer. *"Scavenging of Two-Stroke Diesel Engines."*
5. Watters, Hoover and Franken. *"Designing a Muffler for Small Engines."* Bolt Beranek and Newman Inc.
6. Darrol Stinton. *"The Design of the Aeroplane."* Granada, London.
7. Gordon Jennings. *"Two-Stroke Tuner's Handbook."* HP Books.
8. Peter Demuth. *"Viertakt Modell Motoren."* Neckar Verlaug.
9. David Gierke. Model Airplane News—*"RPM—Real Performance Measurement."* December, 1992; May, August, October and December, 1993.
10. Andy Lennon. Model Airplane News, *"Reducing Drag."* January, 1992.

5
Using tools and jigs

CHAPTER 5

Organize your workshop
by Jim Sandquist

It seems that the fastest-growing segment of the RC hobby is building large-scale aircraft. Even if you aren't into large, gasoline-powered airplanes, you might enjoy building a 1.20-size airplane. So, what's stopping you from stepping up to the plate? Is it that your workshop is too small, or that you don't have the tools to accomplish the task? If you're hesitant because you fear the hangar rash that results from turning around in a small space, maybe I can help you.

THE PROBLEM
In 1991, I bought a ¼-scale Stearman. The plane had two 8-foot wings and a 7-foot fuselage. It was obvious that I had to make better use of my work space. My 9x13-foot shop is probably similar to many of yours. But by carefully planning how to use space, I was able to store a ¼-scale Stearman, a ⅓-scale Extra 300S, a ¼-scale Stinson L-5 and a ⅕-scale P-51—not to mention all my tools, paints, workbench, etc.! I do admit that I seldom have enough room to fully assemble and rig wings. But the family room is adjacent to my shop, so when I have to fit wings to the fuselage, balance the aircraft, or work on some other major assembly step, I head for the family room. With a little patience on

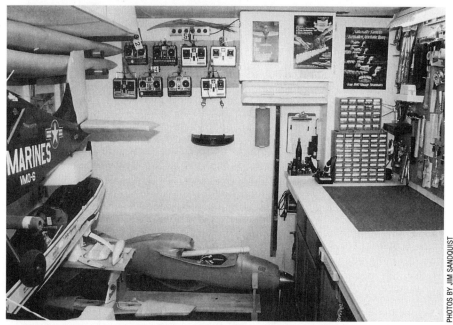
On the left, the planes are stacked floor to ceiling. The pegboard stores transmitters, and there's a power supply above them for convenient charging.

The main work area: tools are easily accessible; power strips have plenty of outlets; and the cabinet on the wall holds a TV and VCR. Under the far end of the counter, you can just see the air compressor.

the part of my wife, I can accomplish my work without much disruption.

THE SOLUTIONS
First, I assembled an unconventional workbench. Instead of using legs to support the bench top, I

USING TOOLS AND JIGS

placed a couple of 30-inch bases on the floor and tied the bench top to them and to the wall. This allowed me to use the space under the bench more efficiently. The bench top is ¾-inch particle board that has a white Formica top, which helps to reflect light and brighten the room. It's 32 inches wide—more than enough space for any ¼-scale model. When I need to build a wing or pin things down, I place a 24x48-inch piece of Builtrite board on top of it.

This family of Dremel tools and accessories is a mainstay. Each of these tools has a specific function.

Above my bench, I have a pegboard that reaches to the ceiling. It keeps all my most needed tools at my fingertips. I have another pegboard that holds my radio transmitters. So that I can do some painting in my shop, I installed a medium-size fan to vent vapors.

Under the bench, I keep a rolling, machinist-type toolbox in which I store most of my hand tools. Also under the bench are my vacuum, drill press, belt sander, bench grinder, scroll saw, band saw and air compressor. Because there is limited floor space, all my power tools are "tabletop," and I put them on the bench only when I absolutely have to. Just below the underside of the bench top is a shelf on which I store wood.

Throw down an inexpensive runner to help minimize foot fatigue, and you'll have everything you need!

TOOLS

Let's talk about some tools that are indispensable to every shop. At one time or another, every modeler has bought the latest and greatest tool, only to have it sit in a drawer. There are lots of valuable tools, but let's take a look at a couple of "must-haves."

- **Dremel tools.** This company has been making power hand tools for modelers for many years. For my first three years modeling, I did not have one. Buying one just always fell low on the priority list. What a mistake! Today Dremel tools are probably the most often-used tools in my shop. I own the cordless Freewheeler, cordless MiniMite and the variable speed model 595. I really like the conven-

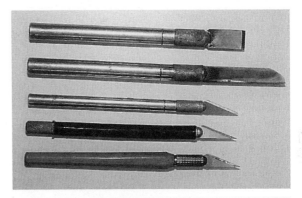
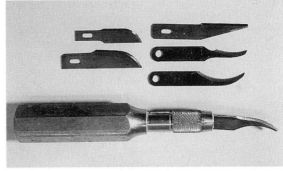

Top: a wide variety of handles, including anti-roll handles, is available. Above: a large handle and wood-cutting blades are excellent for shaping balsa block.

ience of the cordless models. The 2-speed Freewheeler model 850 is generally the first Dremel I grab, however, there are times when the flexibility and power of the model 595 variable speed tool are hard to beat. A slide switch control gives me complete speed control from 5,000 to 30,000rpm. The MiniMite has very low rpm and is great for cutting plastics, polishing and final-shaping wood when the job requires greater control.

Dremel makes a wide variety of tools for cutting and shaping wood, metal, plastics and fiberglass. Companies such as Robart also offer some tools that work with the Dremel motor tools. Robart also has a right-angle attachment for many Dremel tools that allows you to get into tight spaces. If you don't have one of these in your inventory, put it at the top of your shopping list!

WORKSHOP SECRETS 143

CHAPTER 5

- **X-Acto tools.** Every modeler has a no. 11 X-Acto blade and handle in his shop, but you may not know that there are many other useful blades and accessories. I have a variety of handles for different applications. The standard round handle for the

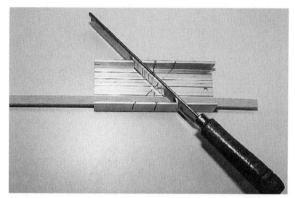
The miter box accessory works great with the 1-inch saw.

no. 11 blade is great, but the other handles with the anti-roll design are also very useful. When I carve or shape wood, the medium-size handles work a bit better. The large handle is not made just for the saw blades; it works very well with any of the wood-carving blades that are available. These wood-carving blades make short work of carving balsa-block nose areas and wheel pants.

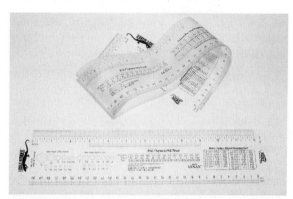
The Rule Bender from TAGS is a highly accurate, completely flexible, Lexan ruler that no shop should be without!

Another must-have cutting tool is the saw blade. Saw blades generally come in ½- and 1-inch versions. The ½-inch blade is a little easier to control, and the 1-inch blade works well in the tabletop miter box. With the miter box, it is quite easy to get consistent, clean cuts. If you have not acquired a power saw, this tool will help you make good joints for gluing.

- **Cutting mats.** Self-healing cutting mats are a nice accessory. Not only will they protect your bench from inadvertent cuts, but they will also prolong the life of your blades. Most of the mats I've seen are marked in inches, which can be helpful when you cut MonoKote to size for covering. These mats are available in many sizes. I have a large 24x36-inch mat and a smaller 8x5-inch mat, which is very portable and useful for cutting smaller sticks. These mats are available at many hobby shops and at most fabric stores.

- **Rulers.** A variety of rulers is a must. Although I never learned the metric system, I like using metric rulers. They are more accurate and somewhat easier to read. I have both flexible and rigid rulers. It's nice to be able to wrap a ruler around the fuse-

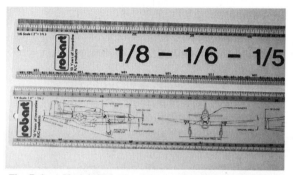
The Robart Modelers Scale makes conversion for scale projects as simple as reading a ruler.

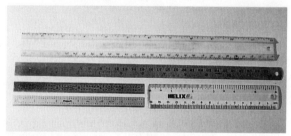
An assortment of rulers makes the job easier.

lage and get a precise measurement. One ruler I recently received is the Rule Bender by TAGS*. This 10 mil Lexan ruler is available in 12- and 18-inch versions, and it's totally flexible. Accuracy is plus or minus .005 inch, it's very easy to read, and it has many useful model references on it. It costs $5.95 or $7.95 and comes with a lifetime guarantee!

For you scale buffs, you must order the Robart Modelers Scale today! This ruler is marked in ⅛, ⅙, ⅕ and ¼ scale. All the conversions have been done! The markings are marked off in highly accurate ¼-inch measurements for the scale you are working in.

There's no doubt that sufficient workshop space and the right tools make the job easier. If you organize your space efficiently and stock it with the proper tools, the building portion of the hobby will be more fun for you! So, the next time you venture into your shop, step back and see where the improvements can be made. Have fun!

Drill bit tips

by Gerry Yarrish

What's the big deal? All you want to do is drill a simple hole. OK; but do you have trouble drilling aluminum motor mounts or thin sheets of metal and plastic? We all want clean, professional results, and no one wants to spend more money than they have to on tools. When you work on your model, do you burn up bits and constantly buy new ones? Do you have trouble with drill bits breaking or walking off the location you've marked for the hole? If so, this refresher course should help you.

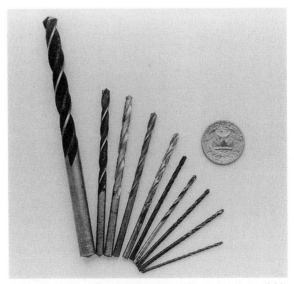

Common twist drills are the most widely used type of bit modelers use. They are best-suited to drilling mild steel and non-ferrous metals such as brass and aluminum.

TYPES OF DRILLS

Some modelers may be surprised at the variety of drill bits available other than just the simple twist drills sold at hobby shops and hardware stores. Here are some of the most useful drills I have on my workbench.

• **Common twist drill.** These drill bits are used primarily for mild steel and non-ferrous metals such as aluminum, brass, magnesium, etc. Modelers often use these bits for all their wood-drilling needs even though other bits do a better job. The greatest incentive for using these is availability; look anywhere tools are sold, and you'll find them. The best way to buy drill bits is in a set, or in a complete package known as a Drill Index Box, but these cost more because they can include anywhere from 8 to 100 bits. The main problem with twist drills is that at the entry and exit holes, they produce burrs and chips, which require additional work to clean the edges of the holes.

• **Brad-point bit.** These drills are much better for drilling into wood, and they don't produce nearly as many burrs as twist drills. The brad-point bit is a hollow ground bit, i.e., it has a negative (almost straight) cutting surface angle. A common twist-drill bit starts cutting from its center point; but a brad-point bit starts cutting from its outer edges. This forces the chips and burrs into the hole and up the drill flutes instead of out the hole.

• **Forstner bit.** To me, these drill bits are like gold. They are expensive (more than double the cost of common twist drill bits), but the fine results you get more than justify the cost. This bit is a combination fly cutter and brad-point bit all in one. It removes wood much like a milling bit removes metal, and it doesn't produce large chips or burrs. Also, this bit is much less prone to walking off center, and it's much stiffer and doesn't bend when in use. For hard balsa, plywood and thin sheets of plywood, fiberglass and plastic, it can't be beat!

• **Combination drill and countersink.** These are best for thick (¼ inch and thicker) metals, and they produce a countersunk hole that guides and supports larger-diameter bits used to drill the finished

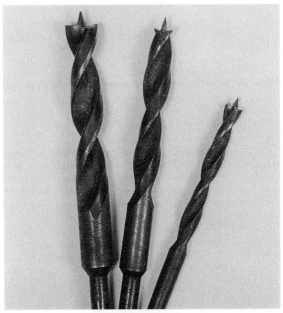

Brad-point bits are much better suited to drilling wood than the common twist drill bit. The brad point bit starts cutting material at its outer edges and produces a clean hole in balsa and pine.

CHAPTER 5

The Forstner bit is like gold! It produces the most precise hole in wood and comes in many sizes. It acts like a fly cutter and a brad-point bit combined.

hole. These are also used as center drills for lathes when a true center hole needs to be drilled into a round piece of metal stock. These drills are very short and very rigid to minimize runout.

• **Spade bits.** These bits are best suited to industrial drilling in which precision is not key. These bits can be used to quickly remove material from block balsa and pine. To use a spade, you must clamp the wood properly and use it with a drill press to minimize wobble. This is my very last choice for model use.

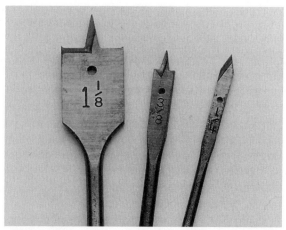

Spade bits are best left in the carpentry shop. They remove material quickly but without much precision. They are my very last choice for model work. And even then, only for non-critical removal of material such as hollowing out a block of balsa.

DRILL USES

• **Metal.** As I said, the common twist drill is best suited to drilling metal. Aluminum landing-gear mounting holes, engine mounts and brass or aluminum mounting tabs are drilling jobs common to modelers. The common twist drill has a cutting angle of 118 degrees and a chisel-edge angle of between 120 and 135 degrees (Figure 1). Actually, a drill cuts away the material being "drilled," and the two front edges (on either side of the chisel-edge center) form a cutting surface similar to a tool bit used on a metal lathe.

To make the best use of a twist drill, take the time to lay out your hole locations precisely, and center-punch them. This will prevent the drill from walking off center and provide a small indentation, which will allow the drill's chisel-edge center to start working before the cutting edges do. For large holes (¼ inch or larger), it is best to first drill the hole with a smaller pilot drill to relieve pressure on the chisel-edge. For harder metals (like mild steel and chromoly steel), it is best to step-drill the holes using larger

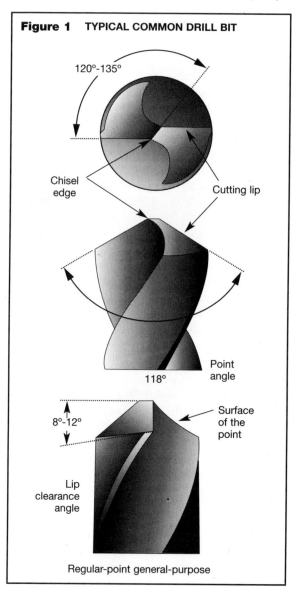

Figure 1 TYPICAL COMMON DRILL BIT

and larger drills to complete the job.

• **Wood.** Whenever possible, drill holes in wooden parts before they are glued to the model's structure. Firewalls, bulkheads and plywood servo trays are easier to handle if you lay them flat on the workbench and drill the holes. For thin materials (1/16 to 1/8 inch), I like to clamp them down to prevent the drill from grabbing and spinning the work piece.

Another way to minimize burrs is to place a scrap piece of wood under the work piece for support. By drilling slightly into the scrap piece, burrs and splinters are greatly reduced.

• **Thin plastic or fiberglass.** To drill holes in very thin materials (0.020 to 0.062 inch), use a contact cement to stick the material to some scrap wood. Use a Forstner bit and high rpm but a low feed. Making scale instrument panels is very easy using this method.

WHAT SIZE?

For properly sized holes, you need to know what drill size to use. Most drills have their size marked on the shank. The size can be expressed fractionally, decimally, or alphabetically. Most drill indexes have labeled drill holders to identify the drills, but you should measure the drill bit before you use it. If you are drilling a hole for a mounting bolt or tapping an engine mount, the size of the hole is critical. The best way to measure the size of a drill is with a micrometer or a Vernier caliper. The absolute minimum tool to use is a drill-and-bolt-size gauge.

CUTTING FLUIDS

For the most part, cutting fluids aren't required for the light drilling work required for model building. But if you decide to use a cutting oil with your drills, be sure to use the correct kind. For aluminum, kerosene works well as does lard paste. For steel, soluble oil and mineral oil work well. In a pinch, WD-40 can be used with aluminum and steel.

So, there you have the "hole" thing—Circular Cutting 101. Using the drill bit that best matches the material you are working with will give you much improved results. You don't have to be a machinist to achieve precise results.

How to use a micrometer and calipers

by Jim Sandquist

Two tools that I found useful when using my lathe were calipers and a micrometer.

USING CALIPERS

The vernier caliper, sometimes called a dial indicator, measures to 1/1,000 inch; it's far more accurate than a common ruler. It comes in different lengths, but the 6-inch version seems to handle most modeling applications and costs around $30. This instrument measures inside dimensions, outside

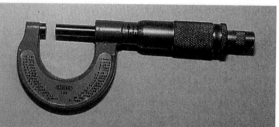

A standard 1-inch micrometer.

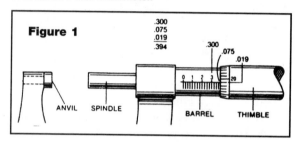

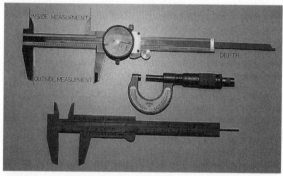

Pictured on the top is a standard dial indicator. Note its three-measurement capabilities—inside, outside and depth. In the middle is the 1-inch micrometer; below is a plastic version of the dial indicator.

dimensions and depth. It has two jaws for measuring; the bar that it travels on is marked off in 1/10 inch, and the dial measures in 1/1,000 inch.

There are several obvious uses for this tool. It can measure the inside of a box fuselage structure, or the outside diameter of wire or tubes, to determine which drill bit to use. It is also useful for marking locations for cutting cross members.

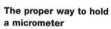

The proper way to hold a micrometer

Making panel lines the same on both sides of a fuselage can be tedious, but with the caliper, it's easy to transpose the lines so that they are the same on each side. Because this tool measures from 1/10 to 1/1,000, it's also easy to find center measurements. Let's say you need to cut in half a balsa sheet that measures 1 11/16-inches wide. Use the caliper to measure the sheet; you'll get a width of 1.6875 inches. Divide that number in two, with a calculator, if necessary, and you get 0.84375 inch. Now dial in that number on the caliper and use it to make cut marks on the wood—quickly and accurately! This may sound a bit foreign, but working in 1/10 inch and using a caliper really is faster and easier than working with a standard ruler. I suggest that you give one a try.

THE MICROMETER

This tool measures outside dimensions on tubing, wire, drills or any other material that will fit within its jaws. Jaw openings generally range from 1 inch to 6 inches. The 1-inch jaw model is large enough for modelers and costs about $15. Reading a micrometer is a little trickier than reading a caliper because:

- each graduation number on the barrel equals one-hundred-thousandth inch (0.100 inch).
- each graduation line on the barrel equals twenty-five-thousandths inch (0.25 inch).
- each graduation line in the thimble equals one thousandth inch (0.001 inch).

For easiest reading, take the highest graduation number on the barrel and add the number of line graduations you see. Then add the number of the thimble-line graduations that coincide with the horizontal line of the barrel (see Figure 1).

These tools have proven to be very helpful in the shop; I think you'll find they provide accurate measurements and are convenient to use. They're available at most good hardware or tool outlets. Good luck!

USING TOOLS AND JIGS

The perfect balance
by George Wilson Jr.

This article on balancing is based on a presentation by Mario Borgatti at the Discover Flying Radio Control Club. With the well-put premise that "A model that is balanced and aligned as the plan indicates will fly as the original did," Mario's talk was aimed at both kit builders and scratch-builders. Mario emphasized that alignment and balancing can be done with inexpensive tools.

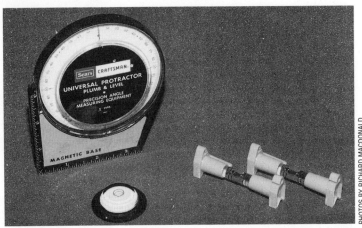

Three types of levels that are useful in balancing models: the plastic line and surface levels are about $2 each at a hardware store, and a smaller version of the Sears dial level is available for less than $10.

NECESSARY TOOLS
The necessary tools for balancing a model are relatively inexpensive. You can buy plastic-bubble line and surface levels at a local hardware store for about $2. To ensure accuracy, check them against a good carpenter's level, and then sand or file them to make them agree. You can also buy a neat 360-degree dial level for less than $10.

Great Planes, Ace RC and Carl Goldberg Models carry balancing stands, or you can build your own. This tool makes adjusting the model's balance point a one-man job and also doubles as an aid in setting the wing's dihedral angle.

Perhaps the most important tool is the simplest: a "level stand" allows you to support the aft end of the fuselage so that the horizontal reference line is level and parallel to your work surface.

These tools can be augmented by blocks and shims that may be necessary to level your building surface. Incidentally, a set of square blocks should be part of every builder's tool collection. Cut them carefully on a table saw. I suggest that you have two ¾x1½x12-inch blocks, two 1½x1½x12-inch blocks, two ¾x¾x12-inch blocks, one 1½x1½x3-inch block and one ¾x¾x3-inch block. The exact dimensions are not important, but when two are called for, they should be alike; cut two from the same length of wood. A coat of sealer (shellac, dope, etc.) will help preserve them.

A CG stand like the one shown here makes fore/aft balancing a one-man job.

If you don't have one, buy a long 3-foot straightedge, or "do it yourself" with a length of aluminum or Plexiglas. Whether you build from scratch or using kits, you'll find many uses for a straightedge.

REFERENCE LINE AND BALANCE POINTS
The fuselage's horizontal reference line and its top and bottom centerlines (CL) should be marked. It's helpful if you mark the bulkhead CLs before you install them; however, the straight line between the firewall's center and the tail-post center is the true CL. You may notice a bow when you compare the CL with the bulkhead CLs, but it should be small enough to be ignored. In any case, when the fin is mounted, it should be parallel to the CL.

Transfer the horizontal reference line from the plan to one fuselage side. Avoid denting the wood by using a very soft pencil to draw the line, or mark several spots and then apply masking tape to show the line.

Mark the proper balance point (from the plan) on the bottom of the fuselage (both sides) and wingtips and on the top and bottom of the wing center. Draw lines from the center marks on the wing outward (perpendicular to the wing's CL). You'll use these later to position the balance stand and to check the wing's position with respect to the fuselage.

WORKSHOP SECRETS 149

BALANCE

Fore and aft balance is the most important balance adjustment. If the model balances too far aft, it will be difficult—if not impossible—to control. If it balances too far forward, it will be "over-stable" and require large amounts of control to maneuver it. Trial-balance the model early in the building process when you can easily move equipment such as batteries and servos—even the engine/motor or wing positions. Balancing the model with extra weight may be necessary, but it should be avoided if possible.

The first step in fore/aft balancing is to level your work surface by placing a level on it and shimming under it as necessary. Level it twice at right angles to each other, or use a round level, which will show levelness in all directions.

Set the model with its flying surfaces, RC equipment, engine, landing gear, etc. temporarily in place on the leveled work surface. Mount a level on the fuselage parallel to the reference line. Tape it on, or find a spot to set it that is parallel to the reference line. Then pin the aft end of the fuselage to the "level stand" when the reference line is horizontal (level). This is the attitude that the model will be in when it is flying "straight and level." The level stand will be used later when the flying surfaces are set to their proper angles.

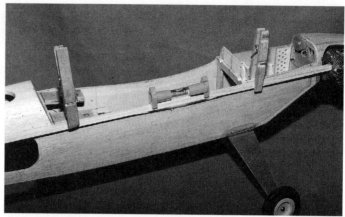
Mario shows the horizontal reference line. This line should be horizontal during level flight and must be set horizontal during the balancing process.

The level stand shown here is a simple tool used to set the horizontal reference line. Sometimes, the simple tools are the best; this one sure beats a stack of blocks!

Lift the model with your fingers under the balance lines marked on the bottom of the wing. With a bit of luck, the reference line will be close to horizontal. To fine-tune the balance, set the model on a "balance stand" with its tips as far from the model's CL as practical and on or near the balance line. Move the model fore and aft until it balances. Shift or add weight as necessary to achieve balance at the balance marks. After the model has been covered and finished, a final fore/aft balance check should be made and the balance adjusted if necessary.

Left/right balance is relatively easy to obtain. It corrects for equipment that is off-center (typically, the muffler). Balance the model on the propeller shaft and the tail post, adding weight as necessary at the wingtips.

If balance weight is needed, always add it as far as possible from the model's center (CG), where it will have maximum leverage and the smallest amount of weight will be required. A heavy propeller nut (available from Harry Higley) or spinner weights (available from Great Planes) are useful if nose weight is needed. If more nose weight is required, place it as near to the nose as possible. Lead flashing from a building-supply house can be shaped and is easy to cut with scissors. Attach it with screws or double-sided tape.

CONCLUSION

The foregoing should not be interpreted to mean that a model must be micro-balanced to fly well. A ¼ inch of tolerance in the balance point of a .40-size model is allowable and probably won't be noticeable. After the model has been test-flown by a competent pilot and flight trims have been made, you can move the balance point forward or backward to make the model more stable or more responsive to the controls. Do this in small steps. Your first experience in trying to control a tail-heavy model will teach you to be very careful.

USING TOOLS AND JIGS

Get the CG right
by Roy Day

You can bet the author's own-design Polish PZL-P38 balances perfectly!

There's a lot of truth to the old saying, "A nose-heavy airplane may fly poorly, but a tail-heavy airplane may fly only once." Correctly balancing a plane is very important; it is surprising how many crashes are the result of improper balance. You may be able to balance your forgiving, straight-wing (constant-chord) trainer on its spar using your fingertips, but a little more care is required for other configurations. This article will explain how to determine where the CG should be located, how to estimate the CG position before you position the model's components and, finally, how to confirm the final CG location.

DETERMINING MEAN AERODYNAMIC CHORD

The mean aerodynamic chord (MAC) for a straight wing is simply the chord of the wing. A good balance point for most configurations is about 25 percent of the MAC, measured back from the leading edge. This is often where the spar is and where the wing is the thickest. (That's why you can use your fingertips to balance that trainer on its spar.)

The MAC of a tapered wing can either be calculated or determined by a graphical method. Figure 1 is a typical tapered wing, where:

R = root chord (centerline chord)
T = tip chord
$$MAC = \frac{\frac{2}{3}(R^2 + RT + T^2)}{R + T}$$

Mark the calculated MAC on the wing outline where it just touches the leading and trailing edges. Now mark the 25 percent point on the MAC and project it to the root chord (R). Distance Y is how far the CG should be aft of the leading edge on the centerline of the root chord. Balance your plane at this point.

If you're uncomfortable calculating the MAC with the formula, here's a simple graphical procedure. Lay this out on your full-scale plans to get good accuracy. Extend the root chord (R) by the tip chord (T), both at the leading and trailing edges.

See Figure 2. Similarly, lay out the root chord (R) on the tip chord as indicated. Now connect these points with the diagonal lines as shown. The MAC is where the diagonals cross. Draw in the MAC and establish your 25 percent point as before. Project

CHAPTER 5

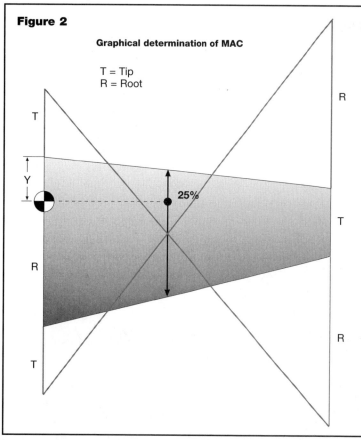

Figure 2
Graphical determination of MAC
T = Tip
R = Root

this point to the root chord (R); this is your desired CG (balance point).

HOW ABOUT BIPLANE WINGS?
In the case of biplanes, the "chord" becomes a combination of the top wing chord plus the stagger. Figures 3A and 3B illustrate how to determine the MAC for the two general classes of biplanes: straight wings with stagger and bent top wings with straight bottom wings. For top and bottom straight wings of equal chord, the effective "chord" is the chord of the top wing plus the stagger (Figure 3A). If there is no stagger, then the top wing chord is all you need to be concerned with, and this case is the same as for a monoplane.

For the case of a swept top wing and a straight bottom wing, use the graphical method as illustrated in Figure 3B and previously discussed.

BALANCING THE AIRPLANE
If at all possible, try to balance the

TABLE 1

Component	Weight (oz.)	Moment arm (in.)*	Moment (oz.-in.)
Props (2)	2	3.25	6.5
Engines (2)	30	5.25	157.5
Engine box (2)	1	7.75	7.75
Engine mounts (2)	2	7	14
Rudder/elevator servos	3	15	45
Aileron servo	1.5	14.25	21.4
Throttle servo	1	15.25	15.25
Receiver battery	3	10	30
Tank	2	12	24
Fuselage	8	17.25	138
Wing and landing gear	30	11.75	352.5
Tail assembly	4	41.75	167
Tail covering	2	41.75	83.5
Fuselage covering and sheeting	6	16	96
Wing covering	3	11.75	35.25
Tailwheel	1	40.5	40.5
Totals	**99.5**	**270.5**	**1,234.15**

$$L = \frac{\text{distance to CG from former F-1 (datum line)*}}{} = \frac{1234.15}{99.5} = 12.40 \text{ inches}$$

Example of calculation of CG for twin-engine airplane.

USING TOOLS AND JIGS

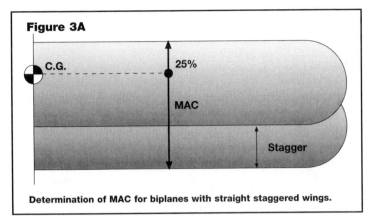

Figure 3A — Determination of MAC for biplanes with straight staggered wings.

CALCULATION OF CG			
Component	Weight	Moment arm	Moment
Engine/prop	W_1	l_1	$W_1 \times l_1$
Cowl	W_2	l_2	$W_2 \times l_2$
Tank	W_3	l_3	$W_3 \times l_3$
Main gear (etc.)	—	—	—
Total weight = W		Total moment = M	

airplane without adding ballast. This can usually be done if you make your preliminary balance check before you cover your airplane and before you finalize the installation of internal components. This preliminary check may indicate a location for the servos and/or battery that will require an access hatch. Even so, this is better than adding ballast weight.

If you're building a new design, or one in which the component layout is not specified or the CG position has not been noted, a calculation of the CG may be very helpful. To do this, you need to weigh all the components of the airplane and estimate their individual CG positions. The large structural pieces (wing, fuselage and tail) can be balanced on the scale as you weigh them. Dense components like the engine, servos and battery have their CGs at roughly the center of their side view. List all the components with their weights and their distances (moment arms) from a common datum line. This is illustrated in Figure 4 and Table 1.

The overall airplane CG position measured from the datum line is the sum of all the component moments divided by the total weight of the model.

$$L = \frac{M}{W} = \text{the distance of the CG from the datum line (Figure 4)}$$

Now, if the CG doesn't come out where you want it (on that 25 percent MAC point), you can reposition batteries and servos on paper and calculate the airplane CG again. This is a lot easier than physically relocating components or having to add ballast. If it becomes absolutely necessary to add ballast, you can calculate the change in CG by adding a given weight of ballast; just treat the ballast weight as another component.

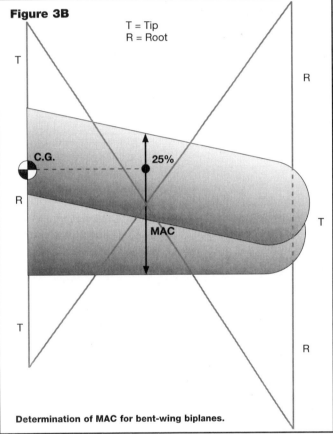

Figure 3B — Determination of MAC for bent-wing biplanes. T = Tip, R = Root.

For my recent original twin-engine design, it was necessary to calculate the CG before installing the components because it had a very cramped fuselage. I needed to position the tank near the plane CG and then find a space for the servos and battery. Table 2 shows my list of components and calculation of the CG. It is important to account for all items. Estimates have to be made for covering, if it has not already been applied.

Of course, calculating the CG must never replace actually balancing the airplane, but it is very useful for scratch-building—particularly if the configuration isn't standard. There are several methods for balancing your airplane, from sus-

WORKSHOP SECRETS **153**

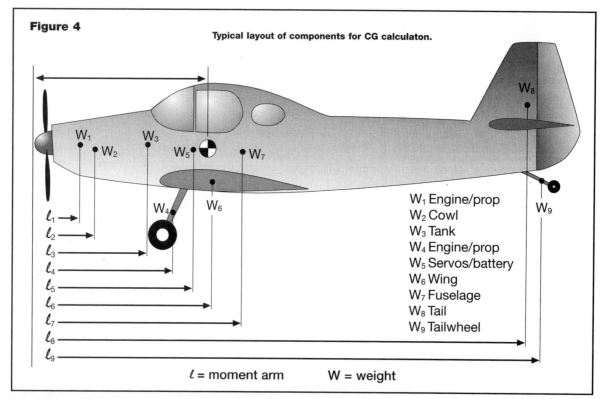

Figure 4 — Typical layout of components for CG calculaton.

W₁ Engine/prop
W₂ Cowl
W₃ Tank
W₄ Engine/prop
W₅ Servos/battery
W₆ Wing
W₇ Fuselage
W₈ Tail
W₉ Tailwheel

ℓ = moment arm   W = weight

pending it to balancing it on supports under the wing. There's even a commercially available CG machine from Great Planes. Use whichever method best suits you, but let balancing the airplane be one of your final checks before that first flight. Now that you know where the CG ought to be and how to determine where it actually is, you can expect that first flight to be successful.

USING TOOLS AND JIGS

Build a jig for Great Planes' Slot Machine
by John Tanzer

I had used my Great Planes Slot Machine before I read the review in the April '99 issue, and I also thought it was the greatest thing since sliced bread. I had one complaint, though: while cutting slots in my Lazy Bee's ⅛-inch-thick stab and elevator, I sometimes didn't get the tool centered just right, so I decided to make a jig for it (it's similar to an adjustable fence on a table saw). What I came up with is a simple tool made of ply and balsa.

The table is made of ⅛-inch ply, the base plate is ¼-inch ply, and two 2-56 rods with nuts and washers allow adjustment. The

After you attach the jig to the Slot Machine, tighten the nuts on the 2-56 rods to lock the jig into place.

Note that the distance between the blade and the table can be changed.

Bring on the hinging surfaces!

fence is ⅜-inch balsa CA'd to the table to form a rigid, 90-degree angle. This homemade jig can be adjusted from 0 to ⅜ inch, and you can use it with the Slot Machine to cut centered slots in ⅛- to ¾-inch balsa. The jig will not work on a tapered surface unless a shim is used to provide a constant thickness. Using the jig will cause the loss of ⅛ inch in the depth of the cut, but you can remove it from the Slot Machine and finish the cut to full depth.

To use the jig, use the Slot Machine's center marking tool to find the center of the surface, adjust the fence so the blade is centered, lock the table and cut the slots. They will be exactly centered.

Now my Slot Machine is perfect. Take the time to make an adjustable fence for your Slot Machine; you won't regret it.

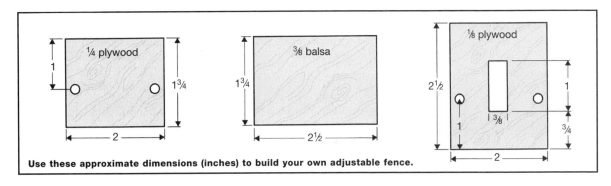
Use these approximate dimensions (inches) to build your own adjustable fence.

CHAPTER 5

Make an inexpensive wire soldering jig
by Joe Beshar

I realized that I needed a third (and possibly a fourth!) hand as I attempted to solder a neat, braided, electric wire splice. How do you hold the skinned wire ends together while holding the soldering iron close enough for the solder to melt and solidify into a neat, strong joint? Twisting the wires doesn't work; the ends unravel. How about tinning the wire ends with solder first? This is also frustrating because the wires must be held firmly to apply the solder with adequate heat and prevent the insulation from melting. Having someone else hold the wire doesn't work either because the human nervous system isn't capable of suspending all movement. I've tried using a holding tree with caterpillar-type clips, but I could never get the clips positioned firmly because they don't have enough "pinching power."

A HELPING "HAND"

To alleviate these problems, I designed the Third Hand for 12-gauge and thicker wire and the Mini Third Hand for the smaller wires used on radio component leads. I've found the Third Hand essential in wiring electric-powered models and radio components. It allows me to connect and adjust wire lengths to make strong joints, ensure minimum electrical resistance and apply heat-shrink tubing easily. The cost of building the Third Hand and Mini Third Hand is nominal. I paid less than $4 for two toenail clippers and $3 for two fingernail clippers.

To make your own Third Hand, remove the pin that holds the arm to the clipper, and discard the pin and arm. Cut the base as shown in Figure 1 then, using the photo at left as a guide, assemble the parts.

Solder the heads of the ¾-inch-long, 6-32 screws to the bottom clipper blades in each side of the Hand. Then turn the knurled nut until the sharp edges of each clipper touch. Run a cutoff wheel mounted in a Dremel tool between the blade edges to dull them (see Figure 2). Now your Third Hand is ready to use.

SOLDERING

Strip the wire ends as desired, and string heat-shrink tubing along the wires and away from the stripped ends. For stability, clamp the Third Hand into a vise or onto a table. Clamp the stripped wire

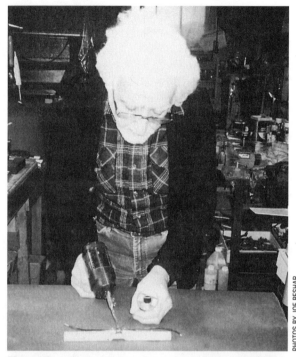

The author demonstrates the convenience of having a Third Hand.

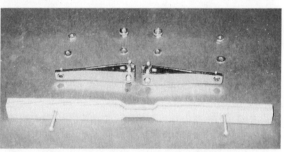

All the parts are in position and ready for assembly.

> **MATERIALS**
>
> **Third Hand**
> - Two toenail clippers
> - Two 6-32, ¾-inch-long machine screws
> - Two 6-32, 1-inch-long machine screws
> - Two 6-32 nuts
> - Four 6-32 washers
> - Two 6-32 knurled terminal nuts
> - One 10x⅜x⅝-inch plywood base
>
> **Mini Third Hand**
> - Two fingernail clippers
> - Four 6-32, ¾-inch-long machine screws
> - Two 6-32 nuts
> - Four 6-32 washers
> - Two 6-32 knurled terminal nuts
> - One 7x⅜x½-inch plywood base

USING TOOLS AND JIGS

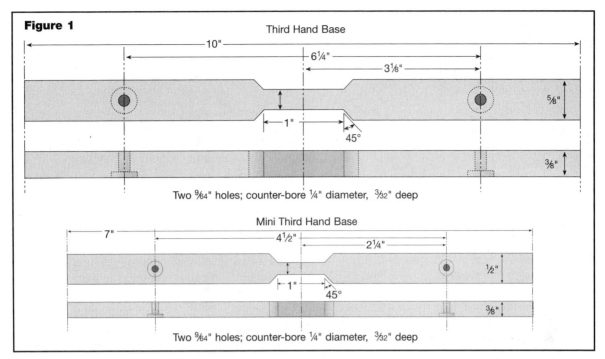

Figure 1

Third Hand Base

Mini Third Hand Base

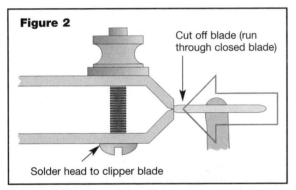

Figure 2 — Cut off blade (run through closed blade); Solder head to clipper blade

ends firmly in the clipper "throats" by turning the knurled nuts. Swing the clamped wire ends out from the base, and tin them with solder. Return both clamped wires into position, placing the tinned ends against one another (squeeze them together with pliers). Solder the joint. Presto! You have a strong, reliable, soldered joint. Remove the soldered splice by unscrewing the knurled nuts. Slide the heat-shrink tubing over the splice, and shrink it with a heat gun.

Making a strong, neat connection is easy with the Third Hand. Try it!

The arms hold the wires firmly in position for soldering.

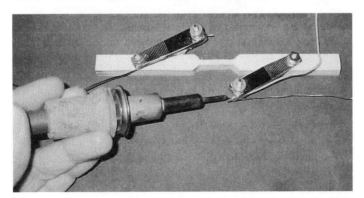

The moveable clipper holds the stripped wire ends away from the Third Hand's base while they are being tinned.

WORKSHOP SECRETS 157

CHAPTER 5

Enlarge 3-views

by Randy Randolph

Scratch-builders are always looking for something new to build, and 3-views from various aviation publications can provide the basis for that next idea. The popularity of PCs and inexpensive hand scanners has made the job of enlarging 3-views not only fun, but also very helpful when creating working plans. A PC, a scanner and a word-processing program that includes graphics will do the trick.

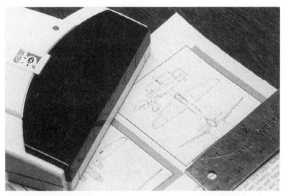

1 You can find 3-view drawings in model and full-scale aviation magazines, advertising brochures, historical and military publications and many history books. Three inches square seems to be a popular format, and it's one that is difficult to enlarge on copy machines.

2 First, scan the 3-view into the computer. Most scanner software offers a variety of scanning options, so choose the one that provides the easiest access to the drawing. Set the scanner for line drawings.

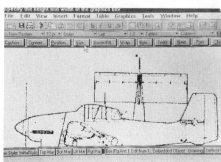

3 Edit the drawing to make the desired component the dominant object of the scan (in this case, the fuselage), and enlarge it to the largest size allowed by the paper size in your printer. Landscape orientation of the page allows the widest part of a standard 8½x11-inch sheet to be used.

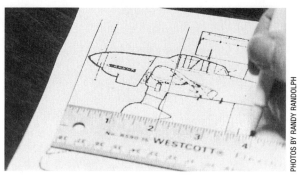

4 Print the scan, then divide it into 2-inch segments. Draw heavy, black, vertical lines at each 2-inch increment. This allows a standard scan width so that all segments can be enlarged the same amount.

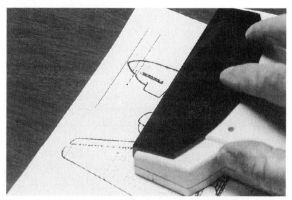

5 Scan each segment separately, and edit each one to the width of the two black lines. Enlarge each segment to a selected size (in this case, 6 inches to produce a drawing three times the size of the print), and print each segment separately.

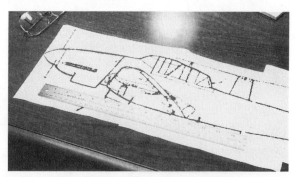

6 Trim each segment to the left black line, and glue them together to produce a finished, enlarged plan view of the original 3-view drawing. The other component parts, such as the wing, stab and front views, can be enlarged the same way.

Enlarge rib sections by Randy Randolph

There has been a lot of excitement about CAD (computer-aided design) and its use with model airplanes—and rightly so. But any computer with word-processing and graphics capabilities and a scanner can easily do one of the things that CAD programs do so well—design scale rib sections for scratch-built projects. It's a simple process, and the photos show the way. Word Perfect 6.1 was used for this demonstration.

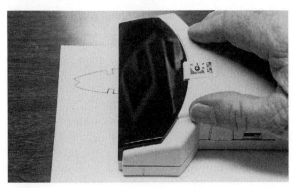

1 First, decide on a rib section that you like (or one that applies), and use your word-processing program to scan it into a graphics box. Rib sections on three-views work just fine.

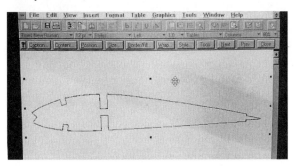

2 When the graphics box has been imported into a document, use the edit box command to make the section larger or smaller; the error margin will be 0.01 inch or less. For a tapered wing, copy the box as many times as necessary and, every time, edit it to the appropriate wing chord.

3 Print out the section, and measure the chord. Compensate for any difference in chord size when you edit the box. Mark the size on each rib. Chords that are up to 13 inches long can be easily scaled with a Landscape format on legal-size paper.

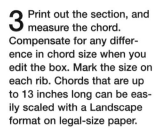

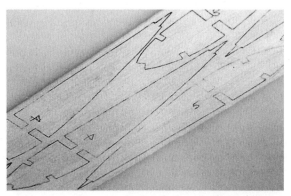

4 If you photocopy the ribs or use a laser printer, the images can be ironed directly onto sheet balsa. Set the iron to the heat that you would use for plastic films.

5 Each printed or copied rib will generally provide two or three good images when ironed onto sheet balsa. If the sections are positioned properly before printing, several rib outlines can be ironed onto a sheet of balsa at one time.

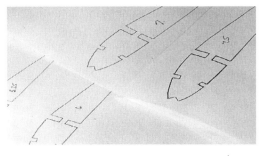

6 For a tapered wing, cut the ribs out of the printed sheet individually. Constant-chord ribs can be cut out of the printed sheet individually, or they can be stacked and band sawed at the same time.

CHAPTER 5

Easy vacuum-forming
by Syd Kelland

Vacuum-forming your own canopies, wingtips, cowls and wheel pants is easier than you might think. Here are some of the secrets that will help you through the process and keep trial and error to a minimum.

I use a wooden vacuum box, my shop vac, a gas barbecue grill, the piece to be molded (called a plug) and some 20-gauge (ga) sheet plastic stapled to a plywood frame. I use the gas grill to heat the plastic because I like to work outside, but a kitchen oven will work just as well. I'm overly cautious when it comes to potentially harmful vapors that may be released when heating plastic.

THE VACUUM BOX

My airtight, 16x18-inch vacuum box is made of 1x4-inch lumber for the sides and ¼-inch-thick plywood for the top and bottom. In the top plywood piece, I drilled a pattern of ⅛-inch-diameter holes roughly ½ inch apart, radiating outward from the center point, and then I joined them with 1/16-inch-deep grooves made with my Dremel tool and a cut-off wheel. I then glued a section of plastic pipe that accepts the hose from my shop vac into a hole on the side of the box. This is where the vacuum comes from.

On top of the box, around its perimeter, I added ½-inch foam weather-stripping tape, which will ensure a good, airtight seal over the entire working surface when I lay the hot plastic on top. Using my ordinary shop vac, I have been able to draw hot plastic down over a 3-inch plug without difficulty.

THE PLUG

Plugs can be made out of almost any material that you like to carve, as long as it can hold its shape under heat and moderate pressure. I carve plugs out of soft wood and then sand them down to the shape I want. I also like to use auto-body filler to add detail and build up any areas on the plug that need extra attention; the putty sets up quickly, is easily shaped with a rasp and can be sanded to a nice, smooth finish. If you make a plug out of foam, you will have to cover it with at least ¼ inch of plaster or body filler to protect it from the hot plastic.

Coat the plug with a thin oil film, which will work as a release agent. Do not use wax as a mold-release agent because the wax will stick to your plug like glue; trust me on this.

THE PLASTIC

Check the "Yellow Pages" for a local plastics supplier; you'll need 20ga styrene sheets (the least expensive and easiest to work with), calendar vinyl and PET-G, ABS, TXP, or any other material suitable for thermal forming. Sheets typically measure 3x6 feet and cost $5 to $20 each, depending on the type of material. Styrene is white, and PET-G and TXP are clear and more suitable for canopies. (This is the kind of rigid, plastic sheet you see formed around consumer electronics products.)

I rough-cut the plastic sheet to the same size as the top of the vacuum box. By doing this, I ensure that there is a good, airtight seal between the hot plastic sheet and the weather-stripping around the vacuum box.

FIRE UP THE BARBECUE

To vacuum-form successfully, it's important to heat your sheet plastic as evenly and as thoroughly as possible. To achieve an even heat distribution on my gas grill, I cover the entire area just above the fake briquettes with a piece of ⅛-inch-thick sheet metal.

Another tip is to make sure the perimeter of the sheet is secured to a wooden frame. I staple the sheet to a piece of ½-inch-thick plywood that's the same size as the top of the vacuum box and has a large cutout in its center. If the plastic is not secured to this frame, it will be deformed (warped) when it's heated. I place the frame on top of a supporting piece of ¼-inch-thick plywood that has a single sheet of paper on top. I put this assembly on the upper warming rack of the grill and then close the lid. The supporting sheet of plywood prevents the warm, sagging plastic from touching any metal inside the grill; if it touched, it would surely melt and make one heck

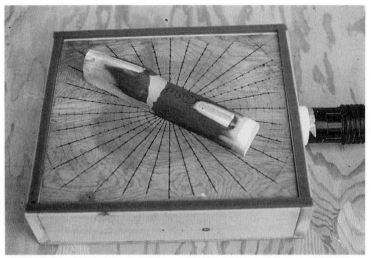

The plug is in position on top of the wooden vacuum box. The shop vac is connected through a hole on its side, and the weather-stripping ensures a good, airtight seal. Note the pattern of holes drilled in the top of the 16x18-inch vacuum box.

The plastic "sandwich" on the upper rack of the gas grill. The sheet metal on the grill ensures a more even heat distribution. I used a staple gun to secure the clear plastic sheet to the plywood frame.

of a mess! The paper prevents the hot plastic from melting onto the supporting plywood sheet.

I keep the temperature inside the grill at around 375 degrees Fahrenheit and set my timer for 3 to 5 minutes as soon as I close the grill's lid. This is not an exact science, but by using a fairly constant temperature, I can vary the "grilling time" for different types and thicknesses of plastic sheet. After 3 to 5 minutes in the grill, the plastic sheet is quite soft and ready to be quickly placed on top of the plug.

Place the plug onto the center of the vacuum box and attach the shop vac. Apply firm pressure to the frame so that it is sealed tightly against the weather-stripping, and then turn on the vac. (A foot switch or a helper comes in handy here.)

CHAPTER 5

Another clear canopy and forward fuselage section is vacuum-formed over the plug.

After just 2 or 3 seconds, the hot plastic will be sucked down tightly around the plug; in another 20 seconds or so, the plastic sheet will have cooled enough to keep its shape.

Turn the shop vac off and remove the finished product from the plywood frame. Trim off the excess material, and that's all there is to it! You could easily make 20 or 30 duplicate canopies in an afternoon using this method.

One of these days, I'll make a larger vacuum box: maybe one that's big enough for a fuselage, but then I would have to figure out a more suitable way to heat up a really large sheet of plastic. Say … pizza ovens are pretty big, aren't they?!

USING TOOLS AND JIGS

Build an inexpensive field stand
by John Gorham

I have been modeling for more than 60 years—and flying RC powered planes and helicopters for more than 25 years—and have loved every minute of it. But as I've gotten older, it has become not only very uncomfortable to kneel, squat and bend over to work on my planes at the field, but also difficult and sometimes painful to stand up again. Now, at 73, I finally decided that it was time to introduce a little more comfort to the hobby. First, I listed requirements for a stand that would make it easier to start, park and service RC models.

The primary requirements were that my flight table take up minimal space when disassembled and not restrict the number of planes I can carry in my vehicle. I also wanted it to be quick and easy to set up at the field. And so, the concept of my portable tabletop field stand was born.

The major design requirements were that the stand:
• Fit vertically behind a car's front seats and not reduce the amount of precious space available for the models.
• Not be thicker than about 4 inches when folded or longer than 46 inches to accomplish the above.
• Be easily and quickly assembled and disassembled at the field.
• Have a strong and safe built-in restraint to hold the model while the engine is started and adjusted.
• Be high enough to allow the pilot to stand upright while starting the engine, adjusting it, or doing repairs.
• Provide readily accessible storage space for fuel and starting equipment and a flat surface for repairing or adjusting the model.
• Have a fuelproof top that can easily be cleaned at the field.

MATERIALS

My search for materials took me to Home Depot. On the way out, I saw a stack of molded, folding, single, sawhorse elements by ZAG for about $20 each. They were perfect. They're sturdy and light and provide an integral tray for the field equipment. They are about 2 inches thick when folded. The only problem was that, because it's a single saw-horse unit, the top surface is only about 2 inches wide—hardly enough for an RC model!

All I needed was a tabletop and the means to mount it rigidly on the stand. I decided on a lami-

The ZAG sawhorse that forms the base of the field stand. These sawhorses are light, fold easily and cost only $20 at Home Depot.

nated framework design. I bought two 10-foot lengths of 2x1-inch pine for $1.50 each and two 2x4-foot sheets of ⅛-inch-thick hardboard for $2.50 each. For the restraints, I purchased two metal flanges (used to anchor 1-inch metal pipe to a wooden surface) and two 12-inch lengths of 1-inch-diameter plastic pipe threaded at both ends. These pipes were then covered with 1-inch-i.d. foam-rubber insulation tubing (see Figure 1).

I selected 40x18 inches as the size of the top; this was large enough for my current models, and it fit into our family wagon behind the front seats. The top can be larger or smaller to suit your needs. The final problem was how to fit the top on the stand

CHAPTER 5

so that it would be strong enough for field operations but readily disassembled.

ASSEMBLY

First, the top must be held so that it doesn't slide sideways or fore and aft. The sawhorse stand already had ¼-inch holes in several places on the top, so this problem was easily solved by fitting two headless screws into the underside. These screws lined up with two holes in the stand. Four 14-gauge steel wire braces between the top and the stand prevented the top from teetering and rising off the stand. These can be rapidly fitted and removed. They are anchored by four screws and wing nuts mounted in the stand's legs.

Figure 2 shows how to make the top. The 1x3-inch (actually ¾x2½-inch) pieces of wood are cut as shown and glued to one piece of hardboard. Glue this unit to the other hardboard piece in the same manner. Before you glue on the final piece of hardboard, mark in pencil the spots on the top at which the wood framework will be glued. Bore the holes for the pipe flanges in the positions shown. Counterbore a hole for the flange of the flange mount first to a depth of about ⅜ inch, then continue the hole at a 1-inch diameter all the way through. If you bore the smaller hole first, you'll have a problem. The flange was mounted, inverted, from the bottom, with three ¾-inch wood screws. To fuelproof the top, I covered it with black Naugahyde (I tested the material at the store before I bought it by wiping some fuel on it). A finish of your choice can be used, but the black Naugahyde looks very classy.

Voilà!—a very functional flight stand and table for field operations for less than $30 (with a little construction work). Keep flying!

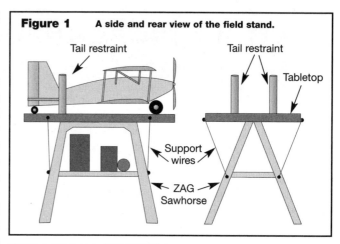
Figure 1   A side and rear view of the field stand.

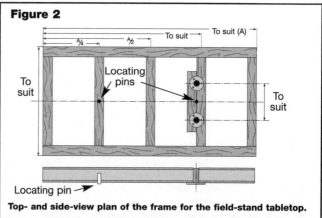
Figure 2
Top- and side-view plan of the frame for the field-stand tabletop.

The two 1-inch-diameter, 12-inch-high threaded plastic pipe restraints rest on top of the field stand. They are screwed into the recessed flanges in the bottom of the tabletop.

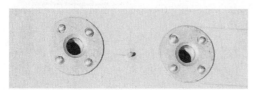
The recessed flanges are installed inverted with ¾-inch wood screws.

The completed assembly with all the components in place. The lower tray holds the toolbox, battery, starter and fuel can. The thin wires support the top when it's placed on the stand.

6 Building projects

CHAPTER 6

Build and install scale cockpits

by Gerry Yarrish

A model—especially one with a large canopy—just isn't complete until there's something to look at inside the cockpit. Many modelers, however, don't consider this extra work necessary. In scale competition, the judges look at the model from a distance, and AMA rules don't require that you decorate the model's "main office." Building a scale cockpit really is a worthwhile effort. The more work I do on a model, the more pleasure it gives me. I build a scale cockpit for me, not the scale judges.

If you have never finished a cockpit, then start with a simple pilot bust and a basic instrument panel. On your next model, add more detail and some depth to the cockpit by lowering the cockpit floor. Before you know it, you'll be building models that have floorboards, rudder pedals and placards, and you'll consider this detail fun and essential.

BASIC LAYOUT

There is no rule that says you have to add scale cockpits to your models, nor are there rules that limit how to get the job done. I keep my projects simple and divide the cockpit area into four or five subassemblies. First, I look at the plans to see if there's enough room in the fuselage to accommodate the cockpit. Usually, it's here that I determine if I have to relocate radio gear, such as servos, pushrods, or batteries (see Figure 1). Then I draw a simple box, or "tub," on the plans that will contain all the detailed parts. Finally, I decide whether the tub can be built outside the fuselage and added later or if it has to be built in place in the fuselage.

Most of the time, I prefer to build the cockpit tub outside the finished model and then simply slip it into place and secure it with screws. This way, the tub can be removed to access other parts of the airframe (for sport flying, you could also remove it before flying and just have it in place for show!). In competition, AMA rules require that no part of the model be removed (except the prop) that changes

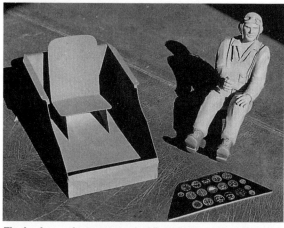

The basics: an instrument panel (installed permanently in the fuselage), the cockpit tub (a single place is shown here) and a scale pilot figure (this one is from Scale Specialties).

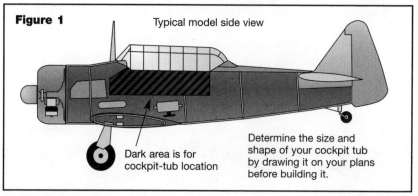

On the left, the planes are stacked floor to ceiling. The pegboard stores transmitters, and there's a power supply above them for convenient charging.

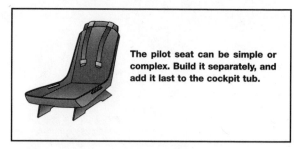

The pilot seat can be simple or complex. Build it separately, and add it last to the cockpit tub.

its scale appearance before flight.

Basically, the cockpit tub can be built as a simple box with its top open. I permanently install the instrument panel in the model because it's less likely to be damaged when it's separate from the tub. I slope the front of the tub slightly, so it can easily slide under the panel when I slip the tub into position. So, what's left?—the floor details, which include the joystick; the yoke and the pushrod details; the floorboards; the rudder and brake pedals; and the seat. Actually, the pilot's seat can be a

166 WORKSHOP SECRETS

BUILDING PROJECTS

major project, but for a first attempt, keep it simple.

If your model is a two-seater, e.g., an AT-6 Texan or a de Havilland Chipmunk, consider installing seat straps or a parachute pack in the back seat to add detail. Next are the side panels and the rear bulkhead and as much detail as you care to add. In cabin models, such as Piper Cubs, Cessnas

I installed the front section of my AT-6 Texan cockpit tub before I painted the fuselage. The panel is installed permanently. Note the throttle quadrant, the side shelf, the fuselage tube detail on the panel side and the joystick, yoke detail and pushrod at the lower end.

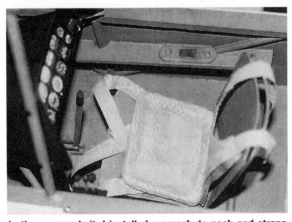

Because the Texan is a two-place version, a rear seat, rollover structure and a rear instrument panel and shroud have also been installed.

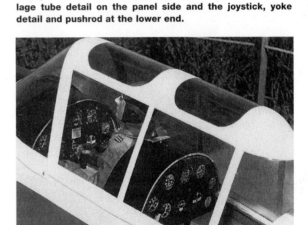

In this half-depth cockpit of my Ohio RC de Havilland Chipmunk, items such as instrument panels, throttle quadrants, pilot figure and shoulder straps make the model look much better.

In the rear cockpit, I installed a parachute pack and straps for added detail. I omitted the rear bulkhead to make room for the retracts' air-bottle installation.

and Stinsons, the entire cockpit is added before the window and door(s) are installed, and the cockpit should usually be added piece by piece after the model has been covered. In either case, making the parts, painting them outside the model and adding them later is the best way to detail your model.

CONSTRUCTION TIPS

Take each part of your cockpit—the floor, the seat, the side panels and the rear bulkhead—and work on each separately. Build each part and paint it, then glue it or, better yet, screw it in place and go on to the next part. Glue the parts together to form the basic cockpit structure. Add more surface details, such as placards, throttle quadrant(s), an 02 hose, a radio headset and lead wires; install the pilot seat(s), and you're finished.

Here are some other tips:
- Use photos and other documentation (available from Bob Banka's Scale Model Research) to duplicate the details you want in your cockpit tub. If you don't have enough detail information, improvise! No one will ever know. A

CHAPTER 6

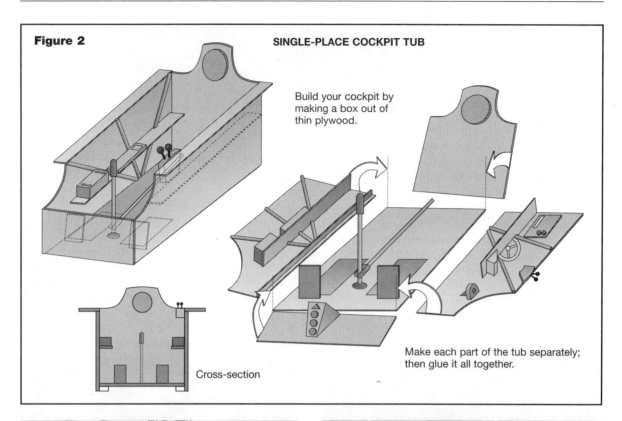

Figure 2 — SINGLE-PLACE COCKPIT TUB

Build your cockpit by making a box out of thin plywood.

Make each part of the tub separately; then glue it all together.

Cross-section

Scale details play a major part in a model's appearance, as shown here in Myron Eister's beautifully appointed 1/3-scale Stinson L-5. Using Vailly Aviation plans, Myron's cockpit detail is a feast for the eyes.

Thanks to the transparent observation window above the cockpit, details such as the radio and fuselage structure are obvious even at a casual glance.

Russian or German fighter will still look great with U.S. instruments on the panel.
- Use lightweight materials; 1/32- or 1/64-inch plywood is strong enough for most cockpit sides and bulkheads.
- Make the floorboard first and, from underneath, add a balsa framework to which the side panels can be glued.
- Add shelves and side details to the side panels before you glue the panels to the floorboard.
- Periodically fit the tub structure in place in the fuselage while you build the tub. This way, you can work out the position of the rear bulkhead so the whole thing slips in and out easily.
- Build a lip about 1/4- to 3/8-inch wide on top of the side panels so you can screw the tub in place from above.
- Add the pilot's seat last.
- Paint all parts first before you add them to the tub. Where items are glued together, remove the paint, when possible, to create a strong bond.
- Don't spend too much time on parts or details that can't be seen from outside the finished model (unless you really want to!)

BUILDING PROJECTS

Not happy with just a beautifully detailed cockpit, Myron included a rear compartment used for transporting wounded soldiers out of combat areas. Even the pilot and wounded soldier are finely detailed.

- Use flat or semigloss paints, e.g., Testor's Master Modeler enamels, and then spray on a flat, clear coat.
- Make seat belts, shoulder harnesses and chute straps by first spraying fiberglass cloth with spray adhesive. Fold the cloth onto itself, and then cut it into strips of the appropriate width. Glue the cloth in place with thin CA, then saturate the cloth with more thin CA to set it into position. Once the cloth has been painted, it looks very realistic (duct tape cut into strips also makes good-looking straps). Buckles and clips can be made out of thin cardboard or plywood and painted silver.
- Finally, add some weathering here and there with smudges of brown paint or splinters of silver paint along the edges and corners. First paint everything silver, then paint the tub with the appropriate interior colors. When the paint is dry, use an X-Acto knife to scrape off some of the finish off so that the silver shows through to simulate weathering.

That's about it. Scale interiors are fun to build, and they're impressive to look at in the finished model. Start small, keep it simple and, above all, have fun!

CHAPTER 6

Make scale cockpit details
by Gerry Yarrish

For many scale modelers, the most problematic chore when building a scale model—fighter, bomber, or civilian aircraft—is decorating the main office—or cockpit. On the flightline, you'll often see a beautiful scale model that has loads of external details, and there's no question that the modeler did his homework. Stepping a little closer, you peer into the cockpit and the illusion ends. Two easy-to-build details that will make your cockpit come to life are the instrument panel and the throttle quadrant. Here's how I made them for my WW II de Havilland Chipmunk trainer.

Good-looking details are not achieved by accident; a lot of documentation needs to be done first. Assemble photos and/or drawings of the cockpit that you plan to build, and try to simplify what you see. The trick isn't to duplicate every single detail but rather, to decide which details are the most important and then focus on them.

1 To make the instrument panel, you'll need instrument faces (I use J'Tec), thin plastic, wood or sheet aluminum (for the panel face) and fly cutters (Forstner bits) that are the same size as the instrument faces' openings (in this case, 5/8 and 1/2 inch). You begin by drawing (on paper) the plans for your panel, which will be located on the model's bulkhead. Make sure that you include all the openings, switches and screw positions.

2 Use spray adhesive to tack down the panel material (I use 0.015-thick prepreg fiberglass sheet) to a piece of wood, and then attach your instrument panel drawing on top of that with more spray adhesive. Carefully drill out each opening (I use a drill press with the fly cutters). Notice that I have marked the positions of all the attachment screws for each of the instrument faces. You can drill the holes out now or you can simply center-punch them and install the screws later.

BUILDING PROJECTS

3 After you have drilled all the instrument face openings, cut the panel to size while it's still tacked to the wood. (I used a band saw with a fine blade and then sanded the edges smooth.) When the panel has been cut, carefully peel it off the wood and clean the adhesive off it with paint thinner or mineral spirits. Now trace the panel outline both onto a backplate that has been made out of 1/16-inch plywood and onto a piece of clear 0.015-inch-thick plastic; cut out these two traced parts. Next, spray paint the panel with an appropriate color. I used flat black "non-skid" paint to obtain a krinkle-like finish.

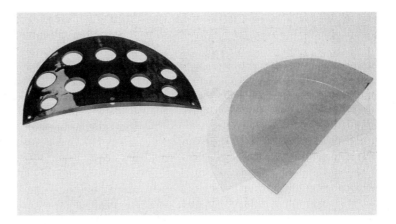

4 Cut out the instrument faces and use ordinary, clear tape to secure them onto the plywood backplate. (Hint: put the panel face on the plywood and trace a pencil line around the instrument holes to mark their positions; then remove the panel face.) Once you're satisfied with their positions, spray the plywood face with a very light coat of spray adhesive and press the clear plastic onto the instrument faces. The adhesive is almost invisible and will not show through the face openings.

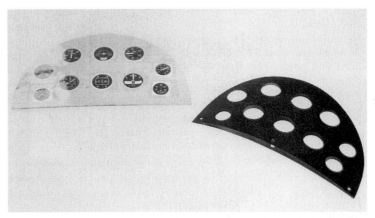

5 Finally, flip the panel over, spray the back with adhesive, then position it over the instrument faces and press it firmly into place. Use a thick book to weight down the assembly for a few minutes, until the adhesive has dried. This ensures that the finished panel will be flat. Now you can add the screws, the switches and other small details. I use Nelson Aircraft Co. subminiature screws as the bezel screws and the panel-mounting screws. Use a thumbtack to punch holes in the panel, and then install the screws with a small screwdriver. The opening photo shows how I used some

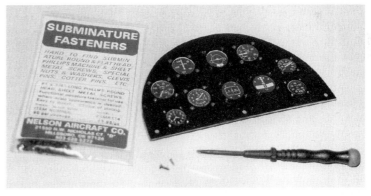

placards that I had made on my Macintosh PC. I made the toggle switches by using the cut-off tips of thumbtacks, which I glued into the slots in the heads of small cap-head screws and painted them silver. Straight pins can also be used as the bezel screws, as shown in the three smaller instruments.

6 I get more comments about the little throttle quadrants that I install in my scale cockpits than I get about any other detail I make. Here are the few parts you need to make a twin-control quadrant: five pieces of 1/32-inch-thick plywood cut to the outside shape of the quadrant; some miniature screws; two small, plastic beads from a child's necklace; and two thin strips of aluminum that have been cut as shown.

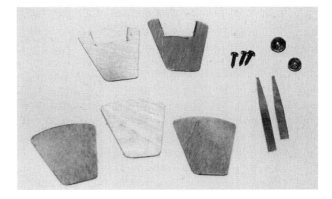

WORKSHOP SECRETS **171**

CHAPTER 6

7 Stack the thin plywood plates as shown, and glue them together with thin CA. Notice the openings formed between the layers by the notches that have been cut into the second and fourth layers. These are the track openings that the "aluminum levers" are glued into. Drill a small hole into the two plastic beads, insert the tapered ends of the aluminum strips into the holes and glue them together with medium CA.

8 Sand the plywood-block quadrant to its final shape and paint it flat black. Drill holes in the corners of the quadrant and install the small Nelson subminiature screws. Paint the plastic beads black. Bend the aluminum levers as shown, and glue them into the track slots with thin CA. The larger knob is the throttle and, depending on the aircraft that you're modeling, the smaller knob can be the mixture-control knob (painted red) or pitch control for the prop (painted green).

It's very easy to make small-scale cockpit details. When they're put together, they make your cockpit come to life and add a lot of realism to your finished model. Apply these simple steps to your next model; you'll be glad you did.

BUILDING PROJECTS

Add power and RC to Robart's foamie F-16
by Nick Ziroli Sr.

The first time I saw the Robart Mfg. F-16 throw glider, I thought about converting it into a powered RC model. Robart produces a series of Top Gun molded Styrofoam throw gliders that are available in department, toy and hobby stores. Besides the F-16, an F-117 Stealth Fighter and a UFO Space Plane are available. These planes all have wingspans of about 18 inches, and the F-16 is more than 25 inches long. No assembly is required, other than inserting the vertical fin in a slot in the top of the fuselage. Each plane comes with a set of colorful stick-on graphics to dress it up. These are rugged models that will take a lot of abuse, but if they do break,

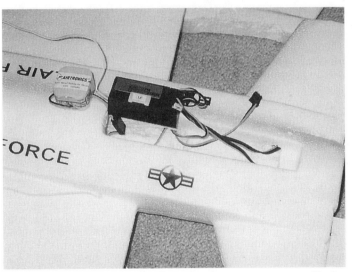

A small Airtronics 4-channel receiver and 100mAh battery are used in the 2-channel F-16; a 275mAh battery is used in the 3-channel version.

the parts can be glued back together with 5-minute epoxy or white glue. With a retail price of $9.95 or less, these Top Gun gliders are a good value, and converting them to power and RC requires only a few hours' work. I used a Cox Golden Bee reed-valve engine on my first F-16 test model. When the needle valve is right, performance is quite lively. For some real excitement, use a Cox Tee Dee .049 to .051 or Norvel .049 to .061. I built a second F-16 and powered it with the Norvel RC .061. This is a great little engine at a reasonable price, and I recommend it for the F-16.

AT THE WORKBENCH

I found that the best approach is to mount the motor, then position the radio system to obtain the correct balance. If done properly, no additional ballast will be needed. Remove the steel disk that's in the nose ballast hole, and cut the nose off the fuselage with a razor saw or hacksaw blade. If you plan to use the Golden Bee, make the cut 2 inches forward of the canopy edge; make it 1½ inches for the Norvel. Make the cut a little long, and sand in 2 degrees of downthrust and 1 to 2 degrees of right thrust. Trace the firewall shape of the nose onto a piece of ⅛-inch-thick aircraft plywood. Cut this out and epoxy it to the nose. A ½- to 1-ounce fuel tank will have to be buried behind the firewall of the Norvel-powered F-16. I used a ½-ounce plastic bottle, but Norvel's ½-ounce fuel-tank engine mount would be perfect for this installation.

The F-16 is controlled by elevons on the stabilizer. This takes two channels and a transmitter that will control a delta or flying-wing model; V-tail mixing will not work. If it is not possible to get the elevon mixing, ¾-inch-wide strip ailerons can be hinged to the trailing edge of the wing. A servo mounted in the fuselage can control them via a Du-Bro ½A aileron linkage (part no. DU-231) or one similar. The best control would probably be a full stabilizer-span elevator of 1x¼-inch balsa tapered to the trailing edge and operated by a servo in the aft end of the fuselage; however, the elevon system is simple,

CHAPTER 6

Small commercial horns and pushrods may be used, or plywood horns and simple wire pushrods, as shown here. Note the switch mounted on the side of the fuselage.

different and works well.

For elevon control, mount the servos with the output shaft about 2½ inches forward of the elevon hinge line. I used two Airtronics 94501 Microlite servos, which weigh about ½ ounce each. Sub-microservos are not really required, although I did use a Maxx sub-micro MX-50 for throttle. Cut a hole for each servo that will be a snug fit. To mount the servos, I simply wrapped their bottom halves and mounting tabs with masking tape then epoxied them into place. A small amount on the sides will hold them securely. When you remove the servos, peel off the tape and they'll be as clean as new.

The elevons should be 1⅛x¼-inch balsa tapered to the trailing edge. Hinge each one with two, ½-inch-wide sheet-plastic hinges or Robart ½A Hinge Points. CA will melt the foam, so use water-based glue such as Pacer Hinge Glue to hold the hinges in place. I made a few 1/16-inch plywood control horns for the elevons, but the small nylon 1/2A ones sold by Goldberg or Du-Bro would work well, too. For pushrods, I used 1/32-inch-diameter wire bent to length and with a Z-bend on each end. Standard pushrods with a Z-bend on the servo end and a small nylon link on the other would allow you to adjust the length.

The slot for the fin was just wide enough for my receiver and battery to fit. I used an Airtronics 92745/72 4-channel receiver and a 100mAh battery pack in the Golden Bee-powered F-16 and a 275mAh battery in the Norvel version. The receiver is about ⅞ inch square. I gouged out the foam to make room for the receiver and battery below and forward of the fin slot, then positioned the receiver and battery to obtain the correct balance point. This should be 13 inches forward of the end of the tailpipe. You will more than likely have to hollow out the fuselage in front of the fin to move the equipment forward for proper balance. Make the opening just large enough for the receiver and battery to fit. You won't need foam rubber padding, as the foam model itself is soft enough to protect the radio. Place the battery in front of the receiver. The lighter the radio system, the farther aft it will have to be positioned. I mounted the battery switch to a piece of 1/32-inch-thick plywood and glued it to the flat side of the fuselage between the wing and stabilizer.

Completed models weigh 11¼ ounces for the Cox-powered model and 12¼ for the Norvel version.

A Norvel .061 RC engine is a perfect match for the F-16. A Maxx MX-50 sub-microservo controls the throttle.

Microservos and receiver such as those offered by FMA Direct could reduce that by an ounce or two.

The antenna on the first model was run out the front of the fin opening, around the leading edge of the wing and bottom of the fuselage, and up and over the opposite wing through a hole in the trailing edge. On the second model, the antenna runs out the fin slot, up through a hole in the top of the fin and trails aft from there. Be sure to tape or pin the fin in place; we lost it on one flight! The glider flew well as long as the fin was still attached to the antenna because the drag was enough to maintain stability. When the fin came off the antenna, the F-16 went into a such a flat spin that it landed with no damage.

Set the elevon throw so the elevator control moves it ½ inch up and ½ inch down. Aileron control moves them ¼ inch each side of neutral. If the ailerons are on the wings, they should travel ½ inch

BUILDING PROJECTS

each way. The elevons or elevators seem to like a touch of up-trim for neutral (1/16 inch at the most). Trimming them at neutral requires moving the balance point aft, and this makes the glider much too sensitive in pitch, even with reduced throw.

FLIGHT TESTS

The first flights on both F-16 models were successful. The only changes I made to the first Cox-powered model were to retrim the elevons and move the balance forward. The Norvel-powered model was set up as described and flew with only minor transmitter trim adjustments. I'm sure I don't have the optimum setup, but it seems to work well for my daredevil flying!

Launching is easy, since all the gliders in the Top Gun series are made to be hand-launched and have finger holes molded into the underside of the fuselage. Hand-launch them at a

The F-16 on a flyby. The model is fast and will perform big loops, rolls and other maneuvers that don't require a rudder. This is not a beginner's model.

good rate of speed and with the wings level.

The Norvel F-16 is very maneuverable. It will perform big loops, rolls, split-S's and many other maneuvers that don't require rudder control. It's also very fast and exciting to fly. My only caution—and this is true of many models—is that the glider won't turn well if the nose is high or in a stalled attitude.

Carelessness on one flight caused a crash that broke one F-16 into four pieces. It could have been repaired at the field and, at home, it took only 15 minutes and some 5-minute epoxy to make it airworthy again. I've had a lot of fun flying these little F-16 hot-rods. How about a club pylon race with these models? It would be interesting to see how the Robart F-117 and UFO gliders adapt to RC. If you're the type of modeler who likes something a little different, give one of the inexpensive F-16s or other Robart Top Gun gliders a try.

CHAPTER 6

Convert a rubber-powered model to a ½A glow
by Randy Randolph

Herr Engineering's J-3 Cub and Ryan ST are some of the most innovative kits to come along in a while. All the balsa parts (there are a number of them in these kits) are laser cut, and they appear to have been machined, sanded and polished! Although these kits have been designed for rubber-powered free-flight, very little effort is required to convert them to 2-channel RC, and the results are excellent. The photos show how to power the J-3 with a Cox Pee Wee .02 and how to install the radio.

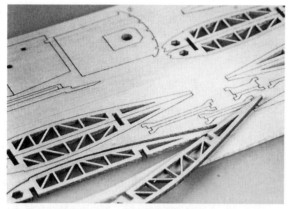

1 Above: this is an example of a laser-cut printed sheet; each part literally falls out of the sheet. The brown edges show where the laser has burned through the wood and left it ultra-smooth and true. The kit includes tissue covering material (more than is necessary), a propeller, rubber for power and associated hardware for the free-flying version.

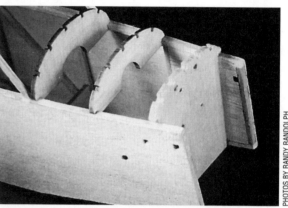

3 Epoxy the firewall ⅜ inch aft of the front of the fuselage sides. Build up the nose block using the parts provided and, after they've been sanded to shape, cut a ¾-inch hole in the center of the thrust line. To eliminate weight and allow a generous amount of clearance for the engine, carve and hollow out the nose block as shown.

2 For RC flying, the fuselage will need a little beef, so build the sides according to plan, then sheet them with 1/32-inch medium balsa. Make the firewall out of ⅛-inch plywood, using former no. 21 as a template. I notched this one, but it isn't really necessary.

4 Glue the nose block to the front of the fuselage, and sheet the top and bottom with 1/16-inch balsa. When the sheeting has been completed, sand the cowl to its final shape. The three holes in either side of the fuselage are for positioning the built-up engine and its scale exhaust system, which is included in the kit—a very nice touch!

BUILDING PROJECTS

5 Follow the instructions for building the wing and the box structure that attaches the wings to the fuselage; this area is strong enough as built. Although the structure is intricate, the remarkably well-made parts result in a very accurate assembly.

6 The plans call for the landing gear to be sewed onto one of the fuselage bulkheads but, because this airplane will weigh more than twice as much as the rubber-powered version, a torsion-type gear should be installed. It adds practically no extra weight and withstands the landing loads much better. The scale shock absorbers were left off to allow the gear to flex on its own. Also visible in this picture is the radio switch and the hatch, which is just aft of the wing struts. The removable wing struts were made by gluing aluminum tubes to the wing and the fuselage at the strut joints. Wire glued to the wing struts hold the struts in place.

7 Once the airplane has been covered and doped, the engine is installed. Paint the inside of the cowl with thinned epoxy glue to protect it from spilled fuel and exhaust residue. The Pee Wee .02 is inverted on its mount, and a hole has been made in the top of the cowl for the needle valve and to allow access to the fuel tank. The engine can be installed through the opening in the front of the cowl.

8 The two microservos, which are mounted on plywood rails and glued to the center cabin longerons, are installed after the receiver and the 220mAh battery pack have been slipped through the hatch into the forward cabin area. The servos can be moved fore and aft to achieve the balance point shown on the plans. The antenna is routed through the aft fuselage just behind the wing and allowed to trail behind.

9 The tail surfaces, built as shown on the plans, are easy to hinge with a sewn figure-8 stitch. The elevator joiner is a 2-inch length of 1/32-inch wire that has been notched into the elevator's leading edge. Elevator and rudder throw shouldn't exceed 1/4 inch on each side of neutral. Because the wing loading is less than 7 ounces per square foot of area, flight performance is good. The J-3 takes off easily with a little forward stick; this gets the tail in the air before rotation. It climbs out very nicely and, like the full-scale Cub, aerobatics is limited. It sure does look great in the air! It's a beautiful kit and a lovely airplane.

WORKSHOP SECRETS **177**

CHAPTER 6

Convert a rubber-powered free-flight to electric RC
by Tom Hunt

Since I started flying RC models in the '70s, the size of receivers, servos and other hardware (except transmitters!) has been slowly decreasing. Now, with true "nano" stuff available, RC model aircraft will soon be kept in shoeboxes in the closet and flown in living rooms! Well, the subject of this article is not quite that small, but it's close!

After spending a few years keeping my eyes and ears open for glow models that I could convert to electric power, I took a break and started a new quest:

The Dumas 30-inch Bearcat is a great choice.

rubber to electric conversions. At the last Toledo show, I picked up the new, 30-inch-span, rubber-powered Dumas F8F Bearcat. It appeared as though it would be a relatively easy task to convert this model to electric RC flight. I decided to give the Bearcat aileron/elevator control because I wanted to keep the dihedral more scale-like. (The model would have needed quite a bit more dihedral to turn properly with rudder, although in hindsight, the model could have supported the extra ½ to ¾ ounce that rudder control would have added.) As long as the Bearcat weighed less than 12 ounces, I wanted to power it with the GD-280 (Titanic Airlines/Graupner Speed 280 motor/gearbox combo imported by Modelair-Tech).

I don't intend to make this a blow-by-blow, stick-by-stick construction article. I will, however, describe my modifications to make this a fine flying, RC electric aircraft. I intended to use as many parts as possible from the basic kit—even some of the scrap wood from the laser-cut sheets!

THE BASIC KIT

The kit is quite complete and comes with laser-cut parts, stripwood, Insignia Blue tissue, dry-transfer decals, vacuum-formed canopy and cowl, wheels, rubber, prop, wire and some miscellaneous hardware. Only glue and dope are needed to build and fly the model as intended. The instructions and plans are clear and well thought out. The laser-cut parts can be removed from the sheet by slicing through two to three tabs per part. Use the plans to identify them first, as no printing is on the wood. Much of the stripwood (the stringers) is very soft. This might be a desirable attribute for an extremely light, rubber-power model, but for a heavier RC model, it's quite a nuisance because it's easily broken during handling. I suggest that you weed out the softer sticks and replace them with some of a harder grade.

THE MODIFICATIONS

• **The wing.** I hate building wings, so I built this first, just to get it out of the way. The wing is built in three sections: the outer panels and a flat center section. I scanned the outer wing section into my computer and drew in an aileron. Slightly larger than scale, the proportions still looked good. The outer ribs would have to be cut back and a sub spar installed just ahead of the aileron hinge line. The aileron itself still uses the remainder of the ribs; however, I duplicated an aft portion of R5 for the inboard side of the aileron. I like top-hinged devices. They're easy to make and install and are completely sealed, so they're very effective. I pinned the leading-edge sheet of the aileron to the

BUILDING PROJECTS

board at a slight angle to provide the gap for the down-aileron travel.

I built the remainder of the wing stock. After I had joined the three sections, I added 1/32, vertical grain, balsa shear webs to the main spar out to the inboard side of the aileron. I didn't make a dihedral brace because the covering material was going to add more strength to the wing than was probably required for the flight loads. I made aileron torque rods out of 1/32 piano wire and 1/16 aluminum tube and installed them in the top of the wing before covering it. At first, it did appear that this light gauge wire (used to keep the weight down) was a bit flimsy. I had some concerns about flutter, but it never materialized. If you can find a suitable tube, you could increase the wire size to .040 or .055 if you'd like. The extra weight should not be a problem.

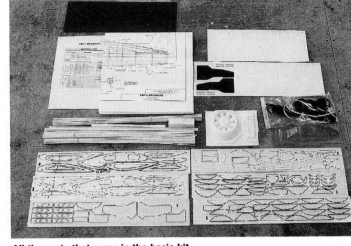

All the parts that come in the basic kit.

I used an FMA S-60 servo for aileron and mounted it very typically

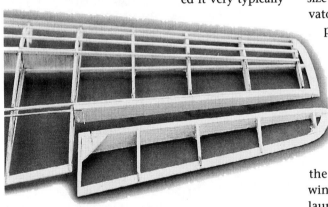

Outer wing panel showing the installation of an aileron.

on a pair of rails supported across the center-section ribs. Using some of the tube that's used to run the torque rod out to the aileron, I flattened the end of about a 3/4-inch-long piece and drilled a 1/32-inch-diameter hole through the flattened end. I then glued this piece to the 1/32 wire torque rod with thick CA, then installed simple, solid, 1/32-inch-diameter wire pushrods with Z-bends at each end and a mid-span V-bend (for adjustment).

• **Empennage.** I built the vertical tail stock. As mentioned earlier, no rudder was used. Both the free-flight and scale horizontal tails are shown on the plan; I compromised and built one somewhere in between using the available laser-cut parts and the hardest 1/8-inch-square balsa I could find in the kit. After I had designed and built scale-size elevator, I made a joiner between the two elevators out of some 0.055 wire that was in the kit. A piece of 1/8-inch dowel with a 1/32 hole drilled through the end functions as the control horn.

• **Fuselage.** What a monster! You could get three or four complete micro radio systems in this fuselage! After building the fuselage per the instructions, I added some 1/32 sheet wood to the inside on which to mount the equipment. I also added some sheet wood to the fuselage sidewall, inside the stringers just aft of the trailing edge (TE) of the wing, to provide a tougher grip area for hand launching. I also framed the section around the cowl as if the cowl were going to be removable (like the rubber versions). I mounted the Titanic Airlines GD-280 motor/gearbox on a plate fastened to the frame around the cowl. I also added some braces as required for this assembly. I decid-

A balsa block is added to the LE to support the wing pin dowel (1/8 hardwood). Note the 1/32 sheet, vertical-grain shear webs.

CHAPTER 6

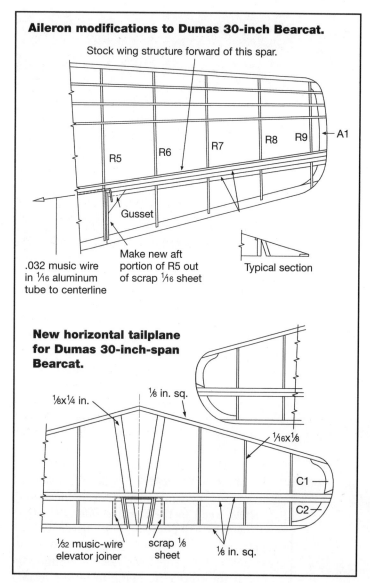

Aileron modifications to Dumas 30-inch Bearcat.
Stock wing structure forward of this spar.
R5, R6, R7, R8, R9, A1
Gusset
.032 music wire in 1/16 aluminum tube to centerline
Make new aft portion of R5 out of scrap 1/16 sheet
Typical section

New horizontal tailplane for Dumas 30-inch-span Bearcat.
1/8x1/4 in.
1/8 in. sq.
1/16x1/8
C1
C2
1/32 music-wire elevator joiner
scrap 1/8 sheet
1/8 in. sq.

Titanic Airlines GD-280 motor/gearbox installed in a vacuum-formed cowl mount; note that additional bracing was done later to stiffen the mount.

ed to retain the cowl/drive with a pair of small rubber bands looped around metal hooks fastened to the cowl, with the opposite end secured to a pair of wooden "hooks" glued to the inside wall of the fuselage, near the wing's leading edge (LE). This arrangement works great! I can access the motor/gearbox by just pulling the cowl forward or by removing the rubber bands. This "soft" mount will also reduce damage to the drive in the event of an accident.

I mounted the wing to the fuselage in much the same manner. A hook on a deck just below the canopy holds one end of the rubber band, and the other end is attached to a hook that's glued into a balsa block in the wing's TE. I strengthened the bulkhead at the wing LE with some sheet wood and made a hole for the 1/8-inch-diameter LE dowel. To install the wing, I simply attach the rubber band to each hook, pull the wing aft enough to slip the LE dowel into the hole and let go! It's simple, effective and shock-absorbent, too! You may need to try a few different rubber bands to get the right tension. I used one double-looped no. 64 on the wing. The cowl is retained by two bands of similar length, double looped but narrower.

After I had shrink-wrapped the elevator servo (an FMA S-60), I simply glued it to the inside sidewall of the model with 5-minute epoxy. The receiver (an FMA Tetra) was installed about mid-wing in the fuselage on a small piece of hook-and-loop fastener. To correctly balance the model, I had to put the battery pack just behind the LE bulkhead. A simple 1/32 wire pushrod attaches the servo to the elevator horn. Again, as I did with the ailerons, I made a mid-span V-bend in the wire for last-minute adjustments. I allowed the ESC (an FMA SC-5) to hang loose inside the fuselage. To keep the weight down, I used 4-pin Deans connectors (two pins for +, two pins for -). Current is only about 3 to 3.5 amps in most cases, and the "doubled-up" Deans connector works well at this level.

FINAL ASSEMBLY

I built the lower part of the fuselage (below the wing) as per instructions, then added miscellaneous sticks to toughen up this area (because it's a belly-whopper!) and to help provide more area for the covering to adhere to. The LE bulkhead should be angled slightly aft to facilitate installing and removing the wing.

I covered the Bearcat in Clearfilm. This relatively light and durable iron-on Mylar is primarily used to laminate documents. It only comes in clear, but it is nearly a third of the weight of most hobby-covering materials. It works at high temperatures, shrinks less than most model brands (a good thing for frail structures) and can be painted on the outside or inside! If painted on the inside (before covering), you must use a fast-drying enamel or acrylic paint so you don't affect its adhesion to the wood or itself. If you paint it on the outside, you only need to clean any latent finger oil off the surface with alcohol.

After I covered the entire model, I cleaned it and painted it with two light coats of Top Flite Insignia Blue LustreKote. If you need to mask sections and add a second color, you may need a base coat of primer. This paint sticks well enough to be handled, but if it is not primed first, pulling off the tape may remove the primary color. I applied the supplied decals a day after painting the model.

THE FINISHED MODEL

The Bearcat, with a complement of 8, 150mAh cells, weighed in at 10.4 ounces ready to fly. I chose this capacity/cell count as a good compromise between weight and duration for the first flight. The wing area computes to just about 170 square inches, yielding a wing loading of a mere 8.5

An FMA S-60 aileron servo is installed in the wing center section. Note the hook that has been installed in the TE block for the wing rubber-band retention scheme.

Equipment installation. Note the rubber band that's loose in the fuselage under the cockpit. The FMA micro Fortress receiver shown here was later swapped for the smaller and lighter FMA Tetra. The 8-cell 150mAh pack is in front of the receiver. The little black "blob" below the battery is the ESC!

ounces per square foot! I have sailplanes with higher wing loadings than that! On the 8-cell pack, the GD-280 spins the supplied 7x6 prop at about 5,000rpm for just about 3.25 amps. That's 26 watts of input power, and a 40 watts-per-pound power loading—actually quite respectable!

The project was a wonderful success. I have taken the Bearcat to all the E-meets I attend here in the Northeast, and it always draws a crowd. It's nice to see that the world of rubber models now has the potential to be converted to electric-power, RC flight. It appears that only common sense "strengthening" would be required for any of them.

On a calm spring morning, around sunrise, I checked the controls and the CG one last time. I tossed the model firmly into the air, with the nose only slightly elevated. To my surprise, it not only continued to climb, but its climb also became steeper! It climbed to about 150 feet, then I throttled back. I was quite pleased with the speed of this fat little warbird. After inputting some slight down-elevator trim, I began to wring out the little model. It looped tightly with speed but would snap roll out if it entered the loop too slowly. It rolled just like a fighter should: a little off axial. Stall turns (to the left) are easy, even without a rudder. Because it's light and spins a larger prop, the model naturally pulls to the left at the top of the maneuver. Inverted flight requires quite a bit of down-elevator, but some travel is left; the thick, nearly flat-bottom wing is responsible for that. The

A temporary stick shows off the cowl rubber-band mounts and additional motor/gearbox bracing.

CHAPTER 6

Model completely framed and ready for covering.

model stalls with a wing drop (either way), but because of the low wing loading, recovery is quick. Without rudder, I could not do any spins, but the wing-drop stalls do not accelerate into spins if left unattended. The first flight lasted about 4 minutes—not bad on 150mAh cells!

Next time out, I tried a 7-cell, 350mAh pack. The model's weight increased to about 11.3 ounces. I certainly expected the flights to be longer—maybe twice as long—because of the higher capacity (and the lower current), but I was not sure what the extra weight would do to the model. This flight did seem a bit sluggish. I felt, though, that I had to stay at a higher throttle setting to keep the model moving. Flight time was barely over 6 minutes. Was the weight too much, or was the power too low?

On the third flight, I added an eighth cell to the 350mAh pack and cycled it a few times to make sure that all the cells were balanced. The model now weighed 11.6 ounces. Another sunrise session showed that not only was the power back, but also that the increased weight did not adversely affect the flight performance. In fact, the flight lasted an astounding 7½ minutes with a lot of aerobatics and high-speed passes thrown in.

I generally fly the Bearcat with the 8-cell, 150 pack when the winds are light. When the winds pick up, I throw in the 8-cell, 350 pack to help the model better penetrate the wind. The model is quite stable and fun to fly. It can be slowed down quite nicely for landings. I do recommend, though, only nice, soft grass for belly-whopping. I would not call this a park flyer; you need a bit more space than a big backyard or a small, tree-lined ball field to fly in. The Bearcat can be kept close by the experienced pilot in a limited area. Despite its big radial cowl, the model does not slow down that quickly with the power off. Leave yourself a lot of room to land the first few times. My concerns during construction and installation of the soft aileron torque rod never became an issue. Control-surface flutter usually is present in soft, "sloppy" installations; however, at the speed at which the model flies, the very low mass of the aileron does not cause any flutter—even in power-on dives from a split-S. Blow-back could be another problem. This is when the aerodynamic load on the surface is so high that the servo cannot push the device into the flow to provide control. Again, I didn't experience this at the speeds this model is capable of flying. I certainly would not expect this to be the case if you "over-powered" this model with a direct-drive 400 motor, though!

Other Conversion Ideas

Here's a short list of some of the other scale models I think might be suitable for the Titanic Airlines GD-280 power system and micro RC equipment described here.

Guillows ¾-inch scale P-47 Thunderbolt
Guillows ¾-inch-scale F6F Hellcat
Guillows ¾-inch-scale F4U Corsair
Guillows ¾-inch-scale SBD-3 Dauntless
Guillows Cessna 172 Skyhawk
Herr (36-inch) J-3
Herr Ryan ST
Herr Fairchild 24
Herr Pitts Special
Dumas Helldiver
Dumas 30-inch Mr. Mulligan
Hacker Tigermoth

Any other model you can think of that has a wing area of between 150 and 225 square inches should be fine. If you're not into scale, many old-timer, rubber-power, free-flight models and Wakefields might also be appropriate.

BUILDING PROJECTS

Convert a small glow plane to electric
by Joe Beshar

The Shrike is manufactured by Lanier RC and, with a .10 or .15 glow engine, it has proven itself a fun and exciting sport model. But I enjoy the reliability and clean and quiet nature of electric power, so I decided to convert the Shrike. The results were not disappointing. Powered by an AstroFlight Cobalt 05 motor and 7, 800mAh cells, my Shrike-E weighs 32 ounces. With a SonicTronics 7x4.5 prop, this system produces 21 ounces of static thrust—more than adequate for good performance.

Before you build the Shrike-E, it is imperative that you familiarize yourself with the building instructions supplied with the kit. Few changes are required to convert the Shrike to electric, so let's get to it.

BUILDING MODIFICATIONS

Begin by reworking F2 and F3 as shown in Figure 1 F3 is installed upside-down relative to the orientation shown in the kit. Drill and tap the 2x56 holes in F2 as indicated. From ⅛-inch plywood, cut out the motor hold-down plate. Build the model as detailed in the kit instructions, but use the

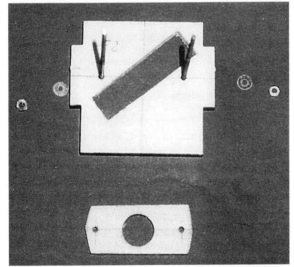

Here's the important conversion assembly. F2 is slotted to permit the brush wires and the rear of the motor shaft to protrude into the area to the rear of the former. The two threaded rods are used to pull the motor hold plate firmly against the front of the motor.

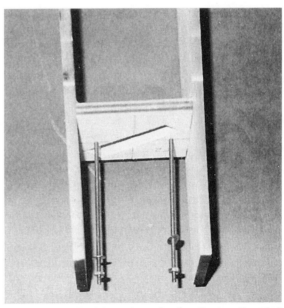

A view of F2 glued into place showing the two threaded rods that will hold the motor in place.

reworked formers. Assemble and glue into place a ³⁄₁₆-inch balsa battery platform as shown in Figure 2, and coat with Balsarite and apply Velcro®-brand fasteners to the platform bottom. When building the fuselage, leave an opening to allow removal of the battery.

I used Airtronics microlite servos for minimum weight and size and a Jeti BEC motor controller in the positions shown in Figure 2. Adjust battery placement to achieve proper balance.

WORKSHOP SECRETS **183**

CHAPTER 6

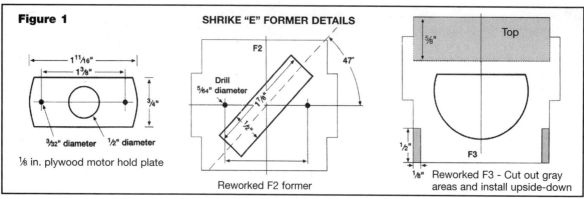

Figure 1 — SHRIKE "E" FORMER DETAILS

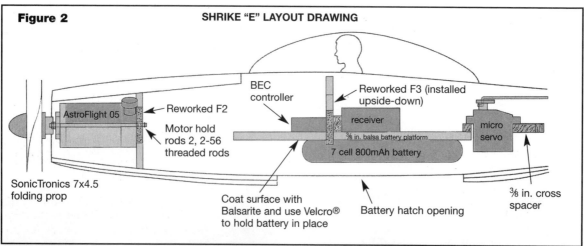

Figure 2 — SHRIKE "E" LAYOUT DRAWING

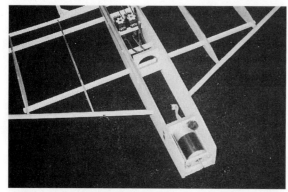

A view of the Shrike-E prior to sheeting. Note the orientation of F3 and the motor mount.

The battery platform is accessible from the bottom and the battery is held in place with Velcro®.

The completed Shrike-E with the top access hatch removed.

I used micafilm on my Shrike-E, as it's light, yet tear-resistant. Apply a strip of ¼x4-inch Velcro® (pin side) to both sides of the fuselage to provide a non-slip grip point while you're launching the aircraft.

Once you've completed your Shrike, check all the electrical circuits and range-check the model with the motor running. Launch the Shike into the wind and enjoy.

184 WORKSHOP SECRETS

BUILDING PROJECTS

Renovate a retired flyer
by Henry Haffke

The author's PT-20A had gathered dust in his workshop for two decades.

I am sure that many of you have an old model lying around that you haven't flown in a long time. It may have been put aside because of interest in a new project, or it may have suffered some damage that you haven't gotten around to repairing yet. These old dust collectors can, in many cases, be turned into a model that's even better than when it was new. Even a model that has suffered extensive damage can be transformed into one that has a lot of life in it. Looking at the damage, you may think it isn't worth repairing. Think again, and look at all the little parts that aren't damaged and could be used to put the model back together again.

I found plans for one of my all-time favorite aircraft, the Ryan ST, in the September '60 *Model Airplane News*. I reworked the free-flight design for 4-channel RC, and as the model was nearing completion, I came across some photos of a military Ryan ST that was designated "PT-20." I liked its silver fuselage, yellow wings and tail surfaces with red, white and blue rudder stripes and stars on the wings. Modeling products of the sixties were not what we have today, and the nearest thing to a silver finish was Aerogloss aluminum dope, which really was more gray than silver. The newest covering on the market, MonoKote, was perfect for the yellow wings. I reworked the model's cowl to hold an Enya .29, thereby turning it into a PT-20A. The 64-inch-span model weighed over 6½ pounds, which is quite light for such a large model. I flew the model extensively and entered it in a few scale meets. Its last flight was made on May 19, 1974.

When I worked for Coverite in 1982, I was involved in developing new products, including Micafilm. Up to this time, most modelers had not been able to finish a metal-looking model unless they used thin aluminum plates for covering.

With a new, larger engine, a few fix-ups and new covering, this renovated PT-20A is ready to take to the skies.

CHAPTER 6

The fuselage is under repair with filler in cracks and clamps holding new longeron in place on the right side. A filler block is used to repair the fin mount.

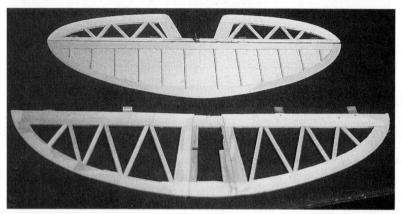

The old stabilizer and the new, resized (scale) stabilizer and elevators ready for covering.

Aluminum paint didn't work well, as it contained aluminum dust that would be rubbed off a beautifully finished model when you handled it. My Ryan PT-20A had been hanging in my shop gathering dust for over 10 years since its last flight. As I had always thought it was a very attractive model, I thought of re-covering it with aluminum-colored Micafilm. This thought stuck in my mind for another 10 years but never got done because of other projects. In the meantime, I was involved with another Coverite product, 21st Century fabric. After covering three models with it, I told myself that I had to refurbish the PT-20A and eventually started late in the winter of 1996.

FRONT-END FIXES

I decided to install a larger engine, a Saito .45. This involved changing the cowl openings and engine and tank mount (built as one unit). I also thought the cowl would be the most difficult part of the model to cover with the aluminum Micafilm. I reworked the engine and tank mount and mounted the engine so the prop was in the same position as it had been with the O.S. .29. The mount spacing didn't require any adjustment, as both crankcases were the same width. I filled the openings where the muffler and needle valve had exited the cowl with balsa blocks and enlarged the cylinder opening to accommodate the larger cylinder on the .45. I refinished these areas and sanded the entire cowl top. I also sanded the lower cowl after filling some small dings that were the result of nose-ups during the model's four years of flying.

When covering (or re-covering) with film, it's important to get the model parts very smooth, as any flaws will show through (especially true with silver covering). I coated the cowl with Balsarite, then I cut a piece of the aluminum Micafilm a little larger than the scale panel, ironed it on with my iron set at 225 degrees and trimmed the edges. Before ironing on the next panel, I brushed a coat of Balsarite on the underside edge of the new piece, where it would overlap the first. After the entire cowl had been covered, I simulated rivet lines with a dressmakers' tracing wheel and used a ½-inch-wide file card as a straightedge. The pressure required depends on the hardness of the balsa underneath. I suggest that you start with less pressure, as it is very easy to go over a line with more pressure. If the line is crooked, you can use your iron to shrink the holes and redo the line.

I sanded down the balsa and ply spinner and covered it with aluminum Micafilm. I painted the back of the propeller blades flat black and covered their fronts with chrome Presto, which is an excellent material to use to trim models and can be applied without heat using your thumb and fingers.

SPRUCING UP

With the front end finished, the next difficult job was the wheel fairings. The fairings had sustained scrapes and other damage and needed a good going-over. I glued small blocks inside the fairings where necessary so their outsides could be reshaped. I filled the dings on the outside with Dapp spackling compound and carved and sanded the parts to their final shape. I brushed on Balsarite, covered each wheel fairing half with four panels of Micafilm and covered the upper leg fairing with one panel. I applied rivet lines as I had done on the cowl.

The next project was to redo the radio compart-

BUILDING PROJECTS

ment hatch cover, which contains the two cockpits. I sanded the hatch and removed the windshields and cockpit coaming. After coating the hatch with Balsarite, I covered it with four pieces of Micafilm and made the seams where panel joints appeared on the real aircraft. I made new windshields and installed a new cockpit coaming. I also made new instrument panels with scale instrument placings, since I had good pictures of both the front and rear cockpit panels. Then I made new aluminum fittings out of aluminum flashing to attach the compression struts and flying wires. I made and installed a cockpit rollover bar, installed a new fuel gauge in front of the front windshield and installed the new windshields and added rivets.

Above: wheel fairings receive balsa repair blocks before refinishing. Right: wheel fairings have been covered with aluminium Micafilm.

FUSELAGE RESTORATION

The next job was to redo the fuselage. First, I cut the tail surfaces free of the main structure. The tail surfaces needed some work, and the top of the fuselage back (where the stab and fin go) was damaged. I cut away the damaged areas and installed balsa blocks where needed. Those of you who remember what happens to a doped and sheeted structure after a good number of years know that it will crack. I repaired the cracks by wicking CA into each one as I held the structure together. I then filled in the uneven areas around the cracks with Dapp spackling compound and sanded everything smooth. The external longeron on the right side of the cockpit area had deteriorated badly due to exhaust residue, so I cut it away, thoroughly cleaned the area and formed a new longeron from a birch dowel.

I sanded down the entire fuselage and applied a coat of Balsarite. Covering the fuselage with Micafilm was easy, as the Ryan's fuselage has no compound curves. (Many 3-views of this aircraft show a curved shape to the rear section in top view. This is incorrect; the rear of the fuselage was a constant cone shape with a straight surface from the cockpit area to the rear.) I cut panels of Micafilm to scale shape and ironed them on, starting at the rear of the fuselage, and then I made the simulated rivets as before. I covered the external longerons with Presto.

Now I had to make a new headrest. This is a simple affair. Cut off a 1-inch-wide balsa block, the bottom of which had been sanded to fit the top curve of the fuselage. On this model, the headrest holds the cockpit section hatch in place. A dowel pin near the rear of the headrest fits into a hole in the top of the fuselage, and a long screw goes through the forward part of the headrest to secure it to the fuselage. The headrest is covered with silver Micafilm, and the headrest "pad" is painted black.

FIXING UP THE FLIGHT SURFACES

The tail surfaces were next. The fin and rudder needed some minor repairs and were then covered with 21st Century fabric. As the original stab and elevators had been repaired a few times, I decided to build a new stab that had scale rib spacing and elevators using a new leading edge and the original (shortened) trailing edges. I covered these parts with 21st Century fabric, hinged the elevators to the stab and fit the stab to the fuselage. Then I added the fin and rudder. The aircraft number "50" was cut out of Coverite graphic trim sheet and applied to the fuselage. I painted the section forward of the front cockpit flat black as an anti-glare panel.

The final job was the wing, which needed only minor repairs to some wingtip scuff marks; these were filled with Dapp spackling compound. The entire wing was sanded very smooth, and after

WORKSHOP SECRETS 187

Closeup showing front-end details.

applying a coat of Balsarite, I ironed on the yellow 21st Century fabric (except at the wing center section, which was covered with silver Micafilm). The red, white and blue wing stars and black "U.S. ARMY" on the bottom of the wing were cut out of 21st Century fabric and ironed into place.

Final detailing involved adding a wing walk of fine sandpaper, which was coated with Balsarite on the underside and then ironed into place, and flying and landing wires and their fairings. The compression struts were made out of streamlined aluminum tube. I fitted and attached the front end to the fuselage, secured the landing-gear fairings to the wires in the wing and bolted the wing to the fuselage.

The Ryan PT20-A shapes into a very striking model that is a real eye-catcher. It's a great feeling to see such a beautiful model back together again. Think twice before you trash the old model that has been hanging around your shop or was damaged the last time you flew it. Rebuilding that old bird can be very gratifying and can save you time and money.

About the Real Aircraft

The Ryan PT-20A is a very interesting and unusual aircraft. It was a later development of the beautiful Ryan ST sport trainer that had been designed as an advanced primary trainer for the Ryan School of Aeronautics. T. Claude Ryan laid out the preliminary

Beautiful flight shot of a PT-20A in the clouds.

design for the ST in 1933. By early 1934, the design had been finalized and construction of the prototype began. The design was very sophisticated and everything worked out beautifully, including the use of the landing-gear design used on the 1932 Gee Bee racers. John Fornasero test-flew the prototype on June 8, 1934. His remarks were very promising: "She's perfectly balanced, handles easily and is the sweetest plane I've ever flown."

In 1936, the first export models were delivered. With the uncertain world affairs of the late '30s, the aviation world changed as most manufacturers shifted their production lines to war-related aircraft. To enter the fast-growing domestic military market, Ryan developed the STA-I for evaluation in the 1939 U.S. Army Air Corps primary trainer competition. Its performance won a 15-plane service evaluation order. An order was placed for 30 models carrying the designation "PT-20," modified slightly with the cockpit load-carrying longerons moved to the outside of the fuselage, larger cockpit openings and rollover posts.

The military soon discovered that servicing the Menasco engines was a nightmare, and they searched for a suitable substitute. Twenty-seven PT-20s were refitted with Kinner R-440-1 engines between October and December of 1940; these aircraft became PT-20As.

A Ryan PT-20 before being converted to a PT-20A.

BUILDING PROJECTS

Add plastic interplane struts
by Roy L. Clough Jr.

Knife-edge flying has fascinated me since, as a youngster, I watched pilot Bob St. Jacques fly his WACO, on edge, the full length of the old Laconia, NH, airport. It just didn't seem possible, but he did it!

Attempting this stunt with RC models reminded me that for years, I had nurtured the idea that biplane interplane curtains would do the trick. I hadn't tried it because I seldom build biplanes. When fellow Winnipesaukee Radio Controllers Club member Chip Richards made me an offer I couldn't refuse—a "Wizard" biplane with a 6-channel radio—I saw a chance to try it the easy way.

For my first attempt, I rebuilt the fin and rudder and enlarged the ailerons to suit my prejudices in such matters. The interplane struts were replaced with chord-width panels of 1/16-inch-thick clear Lexan with strips of black masking tape to fake the struts.

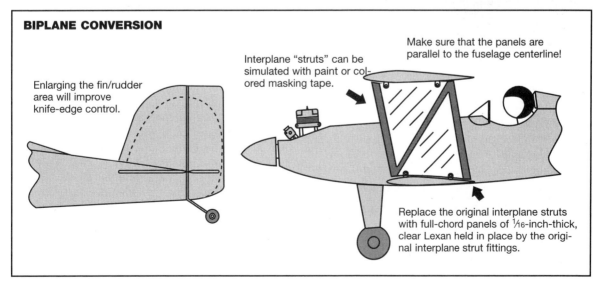

BIPLANE CONVERSION

Enlarging the fin/rudder area will improve knife-edge control.

Interplane "struts" can be simulated with paint or colored masking tape.

Make sure that the panels are parallel to the fuselage centerline!

Replace the original interplane struts with full-chord panels of 1/16-inch-thick, clear Lexan held in place by the original interplane strut fittings.

CHAPTER 6

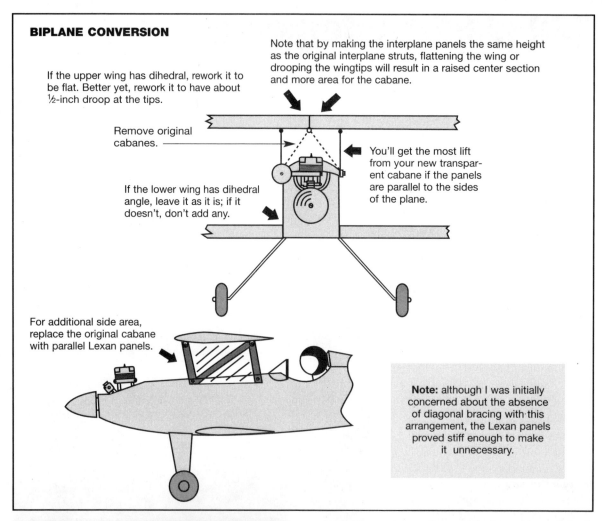

BIPLANE CONVERSION

If the upper wing has dihedral, rework it to be flat. Better yet, rework it to have about ½-inch droop at the tips.

Remove original cabanes.

If the lower wing has dihedral angle, leave it as it is; if it doesn't, don't add any.

Note that by making the interplane panels the same height as the original interplane struts, flattening the wing or drooping the wingtips will result in a raised center section and more area for the cabane.

You'll get the most lift from your new transparent cabane if the panels are parallel to the sides of the plane.

For additional side area, replace the original cabane with parallel Lexan panels.

Note: although I was initially concerned about the absence of diagonal bracing with this arrangement, the Lexan panels proved stiff enough to make it unnecessary.

Test pilot Armand Cote points out the plastic interplane struts.

This added 40 percent more area to the side area of the plane. Knife-edge flight was possible but required a very delicate touch of thumb.

After several fellow members had flown the plane, the consensus was that the dihedral of the upper wing was fighting the aileron input. For my second attempt, I flattened out the dihedral to the point of having slightly drooped wingtips. This raised the center section well above the original, somewhat clunky, plywood cabane. I removed it and made new Lexan panel cabanes, complete with fake struts. The result was a total panel area of 160 square inches, and the flattened wing allowed a reasonably skilled pilot to keep the plane on edge indefinitely.

I'll be modest, but I have to tell you the guys in the club think this is a terrific idea. It's easily adapted to any biplane, and taping on or painting struts over clear plastic is much easier than building up N-strut assemblies.

My reworked Wizard has a 54-inch span and is 42 inches long, and it fits nicely into my station wagon. It could, however, be reduced to a much smaller package by removing the top wing (which is held with eight, 3-48 locknuts) and the lower wing (held with six rubber bands).

One of the neat things about this arrangement is that while standing at a distance, no one ever notices that the interwing strutwork is simply painted onto transparent plastic panels!

Index of Manufacturers

21st Century
distributed by Great Planes.

Aerospace Composite Products
14210 Doolittle Dr., San Leandro, CA 94577; www.acp-composites.com.

AstroFlight Inc.
3311 Beach Ave., Marina del Rey, CA 90292; (310) 821-6242; fax (310) 822-6637; www.astroflight.com.

Balsarite (by Coverite)
distributed by Great Planes.

Bob Banka's Scale Model Research
3114 Yukon Ave., Costa Mesa, CA 92626; (714) 979-8058.

Bob Smith Industries
8060 Morro Rd., Atascadero, CA 93422; (805) 466-1717; fax (805) 466-3683.

Bowman's Hobbies
21069 Susan Carole Dr., Saugus, CA 91350; (805) 296-2952; fax (805) 296-9473.

Carl Goldberg Models
4734 W. Chicago Ave., Chicago, IL 60651; (773) 626-9550

Cox Hobbies
P.O. Box 270, Penrose, CO 81240; (719) 372-6565.

Coverite
distributed by Great Planes.

Deans
10875 Portal Dr., Los Alamitos, CA 90720; (714) 828-6494; fax (714) 828-6252; deansco@earthlink.net.

Diversity Model Products
distributed by New Creations R/C.

Du-Bro Products
P.O. Box 815, Wauconda, IL 60084; (800) 848-9411; fax (847) 526-1604; www.dubro.com.

EZ Lam
distributed by Aerospace Composite Products.

FMA Direct
9607 Dr. Perry Rd., Unit 109, Ijamsville, MD 21754; (800) 343-2934; fax (301)831-8987; www.fmadirect.com.

Goldberg
see Carl Goldberg Models.

Great Planes Model Distributors
2904 Research Rd., P.O. Box 9021, Champaign, IL 61826-9021; (800) 682-8948; fax (217) 398-0008; www.greatplanes.com.

Hangar 9
distributed by Horizon Hobby Inc.

Hayes
distributed by Carl Goldberg Models.

Herr Engineering Corp.
1431 Chaffee Dr., Ste. 3, Titusville, FL 32780; (407) 264-2488; fax (407) 264-4230; www.iflyherr.com.

Hobby Lobby Intl.
5614 Franklin Pike Cir., Brentwood, TN 37027; (615) 373-1444; fax (615) 377-6948.

Horizon Hobby Inc.
4105 Fieldstone Rd., Champaign, IL 61822; (217) 355-9511; www.horizonhobby.com.

JR
4105 Fieldstone Rd., Champaign, IL 61821; (217) 355-9511; www.horizonhobby.com.

J'Tec
164 School St., Daly City, CA 94014; (650) 756-3400.

J&Z Products
25029 S. Vermont Ave., Harbor City, CA 90710; (310) 539-2313.

K&S Engineering
6917 W. 59th St., Chicago, IL 60638; (773) 586-8503.

Koverall
distributed by Sig. Mfg.

Lanier RC
P.O. Box 458, Oakwood, GA 30566; (770) 532-6401; fax (770) 532-2163.

Loctite Corp.
18731 Cranwood Ct., Cleveland, OH 44128; (216) 475-3600; customer service (800) 338-9000.

Maxx Products
815 Oakwood Rd., Unit D, Lake Zurich, IL 60047; (847) 438-2233; fax (847) 438-2898.

Micafilm by Coverite
distributed by Great Planes.

Midwest Products
P.O. Box 564, Hobart, IN 46342-0564; (800) 348-3497 or (219) 942-1134; fax (219) 947-5703.

Modelair-Tech
P.O. Box 1467, Lake Grove, NY 11755-0867; (631) 981-0372; www.modelairtech.com.

Model Research Lab
25108 Marguerite #160, Mission Viejo, CA 92692.

MonoKote
distributed by Great Planes.

Nelson Aircraft Co.
394 S.W. 211th Ave., Aloha, OR 97006; (503) 629-5277; fax (503) 629-5817.

Northeast Aerodynamics Inc.
P.O. Box 208, Methuen, MA 01844; (978) 686-0319.

Norvel
P.O. Box 3459, San Luis Obispo, CA 93403-3459; (800) 665-9575; (805) 547-8360; fax (805) 547-8365; service@norvel.com; www.norvel.com.

Oracover
distributed by Hobby Lobby Intl.

O.S.
distributed by Great Planes; www.osengines.com.

Pacer Technology
9420 Santa Anita Ave., Rancho Cucamonga, CA 91730; (909) 987-0550; (800) 538-3091.

Robart Mfg.
P.O. Box 1247, 625 N. 12th St., St. Charles, IL 60174; (630) 584-7616; fax (630) 584-3712; www.robart.com.

Rocket City R/C Specialties
103 Wholesale Ave. NE, Huntsville, AL 35811; (205) 539-8358.

Sig Mfg. Co. Inc.
P.O. Box 520, Montezuma, IA 50171; (800) 247-5008; (515) 623-5154; fax (515) 623-3922; www.sigmfg.com.

Solartex
distributed by Global Hobby Distributors.

SonicTronics
7865 Mill Rd., Elkins Park, PA, 19027-2796; (215) 635-6520; fax (215) 635-4951.

Superior Aircraft Materials
12020-G Centralia, Hawaiian Gardens, CA 90716; (562) 865-3220; fax (562) 860-0327.

Tekoa: The Center of Design
49380 Skyharbor Way, Aguanga, CA 92536; (909) 763-0464; fax (909) 763-0109.

Testor Corp.
620 Buckbee St., Rockford, IL 61104; (815) 962-6654; fax (815) 962-7401.

The Aeroplane Works
2134 Gilbride Rd., Martinsville, NJ 08836; (908) 356-8557.

Top Flite
distributed by Great Planes.

Ultracote
distributed by Carl Goldberg Models, 4734 W. Chicago Ave., Chicago, IL 60651; (773) 626-9550.

Vailly Aviation
18 Oakdale Ave., Farmingville, NY 11738; after 6:30 pm EST (516) 732-4715.

West System
distributed by Composite Structure Technology, P.O. Box 642, Tehachapi, CA 93581-0642; (805) 822-4162; (805) 822-4162.